Famines and Agrarian Conditions in South India

A Case Study of Rayalaseema in Andhra Pradesh

Famines and Agrarian Conditions in South India

A Case Study of Rayalaseema in Andhra Pradesh

Pratap Kanakalapati

SHARADA

Sharada Publishing House

Delhi

First Published 2025

ISBN: 978-93-83221-63-9
₹ 1750

Published by
SHARADA PUBLISHING HOUSE
111, Vidya Vihar Apartment
Sector-9, Rohini, Delhi-110 085 (INDIA)
Ph.: +91-11-4702 6210 (O), +91-9899028954 (M)
E-mail : books@sharadabooks.com
bansal_sph@rediffmail.com
Website : www.sharadabooks.com

Printed at
Delhi

FOREWORD

Famines in ancient and medieval India, even during the Hindu period of her history, India did not enjoy absolute immunity from famines. But judging from the infrequency of allusions to these calamities in the ancient Sanskrit works and from the testimony of foreign travellers, it would not be incorrect to say that famines were an exceptional occurrence in ancient India. When they did occur, adequate relief measures were undertaken by the state. Kautilya in his *Arthashastra* mentions among other relief measures the revision of taxes, emigration, the granting of money and grain from state funds, construction of artificial lakes, tanks, wells, etc. and the importation of grain from other places. During the Mohammendan period, there were several famines and scarcities, the most severe occurring in the reigns of Mohammad Tughlak, Akbar, Shah Jahan and Aurangzeb. In spite of vigorous measures adopted by the emperors the famines took a heavy toll of life.

During the rule of the East India Company India suffered in one part or another from 12 famines and 4 severe scarcities. The first of these was the dreadful Bengal Famine of 1769-70 which claimed a third of the population of the province. No relief measures worth the name were undertaken by the state. Rather, the Company servants made large profits by buying up rice and retailing it at high prices. The years 1781 and 1782 were years of scarcity in Madras; in 1784 a severe famine afflicted the whole of Northern India. During the Madras famine of 1792 the state opened relief works for the famine-stricken. The Famine Commission of 1880 noted that "till the end of the 18th century the position of the British in India was not such as either to create any sense of general obligation to give relief or to supply sufficient means of affording it".

During the 1803 famine in North-Western Provinces and Oudh (now Uttar Pradesh) the state granted remissions of the revenue gave loans and advances to landowners, offered a bounty on all grain imported into Benares, Allahabad, Kanpur and Fatehgarh. The Guntur Famine of 1833 took away a heavy toll of life; in that 2 lakh persons died out of a total population of 5 lakhs. In 1837 there was a severe famine in Upper India. Public works were opened at several centres. However, the work of relieving the helpless and the infirm was left in the hands of the charitable public.

Under the East India Company no attempt was made to formulate any general system of famine relief or prevention. However, the provincial governments and district officers tried various experiments to afford relief to famine-stricken areas such as the storage of grain by the Government, penalties on hoarding, bounties on imports, advancing loans for sinking of wells etc.

The transfer of power from the Company to the Crown and the economic developments of the latter half of the nineteenth century like the extension of railways and other means of communications and transport, growth of overseas trade changed the complexion of the problem. The state also realised its responsibility for expansion of irrigation facilities, enactment of agrarian legislation and adoption of preventive measures as well as formulation of famine relief policy to meet possible famines.

Under the Crown there were ten severe famines besides a large number of scarcities. The first famine occurred in 1860-61 in the area between Delhi and Agra. This was the first occasion on which poor-houses were used as a means of affording relief and it was the first time when the authorities thought fit to enquire into the causes, area and intensity of the famine as well as took measures to cope with the distress. Colonel Baird Smith was deputed for this purpose but his report did not lead t any formulation of general principles of relief.

The drought of 1865 followed by a famine the following by affected Orissa, Madras, Northern Bengal and Bihar. The calamity was most severe in Orissa, hence the name the 'Orissa Famine.' The Government officers though forewarned took no steps to meet the approach of the calamity and when it came looked helpless. The Government adhered to the principle of free trade and the law of demand and supply. The Government did provide employment to the able bodied, leaving the work of charitable relief to voluntary agency. Since voluntary agency did very little, the famine took a heavy toll of life. It was estimated that 13 lakh persons died in Orissa alone.

The great famine of 1876-78 was perhaps the most grievous calamity experienced since the beginning of the nineteenth century. It affected Madras, Bombay, Uttar Pradesh and the Panjab. The famine affected area was estimated at 257,000 sq. miles with a population of more than 58 millions. Many villages were depopulated and large tracts of territory went out of cultivation. R.C. Dutt has estimated that 5 million persons perished in a single year. The Government made half-hearted efforts to help the famine-stricken. The government famine machinery was inadequate and ineffective and the unwisdom of the policy was amply clear. The Government refused to recognise its responsibility for saving human lives and "the task of saving life, irrespective of cost, is on which is beyond their power to undertake, and that in the interest of the distressed population itself, as well as of the taxpayer generally, The Government of India was bound to adopt precaution against indolence or imposition."

The Government accepted in general the commission's recommendations and steps were taken to find new resources for the creation of a Famine fund to meet extraordinary charges. In 1883 the provisional Famine Code was formulated which formed a guide and a basic for the various Provincial Famine Codes which were subsequently formulated.

Between 1880 and 1896 there were two famines and five scarcities, all of them of more or less local character. The great famine of 1896-97 affected almost every province, though in varying degrees of intensity, the total population affected was estimated at 34 millions. The relief operations were conducted with a fair measure of success except in the Central Provinces where the death rate rose very high. Extensive relief operations were undertaken and in many parts of the country people were relieved in their own homes. The total cost of relief was estimated at 7.27 crores.

Recommendations of MacDonald Commission, Lord Curzon appointed a Commission under the presidency of Sir Anthony MacDonnell. It submitted its report in 1901 in which it summarized accepted principles of relief, suggesting variations wherever necessary. The Commission emphasize the benefits of a policy of "moral strategy", early distribution of advances for purchase of seed and cattle and sinking of temporary wells. It also advocated the appointment of a Famine Commissioner in a province where relief operations were expected to be extensive. It also emphasized enlistment of non-official assistance on a larger scale and preference in particular circumstances of village works to the large public works which had hitherto been the backbone of relief schemes. The Commission also stressed the desirability of better transport facilities, opening of agricultural banks, improvement of irrigation facilities and vigorous measures to foster improved methods of agriculture.

The Great Bengal Famine of 1942-43 took a heavy toll of life.The root cause of the famine lay in a series of crop failures that Bengal experienced from 1938 and in the conditions created by the Second World War. The normal import of rice from Burma stopped and trade and movement of food grains was dislocated because of controls and nearness of Bengal to the theatre of war in the East. This famine might be called "more man-made than an act of God". Man exploited the situation created by Nature and War.

It is evident that famines in India are frequently occurring calamities. The present food situation in the country is a mere extension of the famine problem that has been with us for long. The complexion of the problem has changed as a result of the momentous changes in the political and economic field since 1947, but the basic issue the failure of supply to keep pace with the growing demand for food remains the same. Increasing population coupled with improved standard of consumption of food grains among the poor classes have added to the complexity of the problem. The National Government put 'self-sufficiency' in food grains as the immediate goal of our agricultural policy. But the country even after the successful completion of five Five-year Plans finds itself more distant from that goal today than it was at the beginning of the First Five-year Plan. Now the emphasis has shifted from meeting a famine when it occurs to avoiding its visitation. The "Green Revolution" propped under the leadership of Smt. Indira Gandhi has somewhat lessened the complexity of food problem and the target of self-sufficiency is nearly achieved.

The main focus of this work is on the famine and the agrarian system in general south India and of famines in Rayalaseema region in particular. The first chapter deals with the introduction it includes the area of the study, the main objectives of the book. The second chapter discusses the origin, and historical background of famines, in which a detailed historical evolution has been discussed at length. The third chapter explains the economic deprivation of agrarians, this chapter throws light on the incontinences caused to the agrarians in their day to day life due to sever famines occurred in this region. The fourth chapter probes into the impact of famines on agrarians conditions in Rayalaseema. The policies of the government to compact the situation and the reactions of the civil society has been discussed by means of talking different sample studies in the study area. The fifth chapter focused on various facets and causes of famine and its impact on agrarian system of drought prone area, to which the Rayalaseema region in general, Anantapur district in particular fell as a victim right from the historical period to the present day. The sixth chapter is a conclusive chapter presenting different paradigms of famine issues, social relevance of victims, various view points and observations in the context of famines in Rayalaseema region. In this chapter a detailed account on the recommendations, reports and the reaction of Civil Society was meticulously enumerated to justify the findings and arguments of the researcher.

Prof. A. Satyanarayana (Retd.)
MA, M.Phil (JNU), Ph.D (Heldelberg)
Department of History
Osmania University,
Hyderabad - 500 007 (A.P.)

Contents

Preface

The famine of 1876-78 was ravaging the whole of Southern Peninsula of India (Bombay Presidency, Hyderabad and Mysore Princely states and the Madras Presidency). In the Madras Presidency it affected severely, in fourteen districts out of the twenty-one districts covering in area of 8,000 miles, the distress was felt. It manifested itself in many ways in the country districts by people leaving their home, and wanderers in crowding to *chatnuns* (relief houses), and to large towns; in large crowds congregating around the dwellings of European officers and clamouring for employment or food. The population of Ceded districts was at 39,78,840.

The famine of 1891-92 initially affected nine of the nineteen districts of the Madras Presidency. It was more severe than those of 1854, 1866-67 and 1889-91. But it was of lesser intensity than that of the Dhata (1876-77). The 'Deccan' districts of Anantapur, Kadapa and Kurnool, which are now in Andhra Pradesh were very severely suffered the famine of Khara. Of them, Anantapur district suffered the most and was chosen for the present treatment. The Collector of the district displayed disinterest lethargy and carelessness in running the administration which unfortunately and inordinately delayed the relief works. On this theme, there is no article, no monograph, no research work, not even secondary source material available either in English or in Telugu. This maiden attempt is based on original primary sources culled out of the official repositories of contemporaneous nature. Thus the theme is presented in analytical analysis.

Presidency level comparison, even at the Presidency level, the scanty rain fall, at a comparison of other affected districts shows the magnitude of the problem that was to ensue. Comparison of total rainfall of distressed districts, 1891-92, District and Khara rainfall (in inches): 1. Chengalpattu (29.44), 2. North Arcot (26.36), 3. Coimbatore (21.10), 4. Salem (19.46), 5. Nellore (16.99), 6. Kurnool (14.79), 7. Kadapa (14.17), 8. Anantapur (10.86), 9. Bellary (9.92), such was the indication about the situation in Anantapur District.

The district situation during the Vikruthi year (1890-91), one year before Khara, the district had already suffered famine. Even during Khara (1891-92), the rainfall was rather appalling. The dry season of April-May 1891 received only one-fourth of normal rainfall. The southwest monsoon, the major monsoon of the region, also resulted in sixty percent failure of rainfall. Further, the northeast monsoon which was normally a little more than half of the southwest monsoon also severely failed. To cap all these, even the dry (April-May) season of 1892 was also a severe failure.

The district had an area of 5,275 sq. miles with a population of 7,09,000. Of this, the affected portion was 3,707 sq. miles in area and population of 4,79,000 persons. Of the seven taluks, four taluks namely Dharmavaram, Anantapur, Tadipatri and Gutti suffered substantially. At a few points of time and places, the famine touched the level of Dhata famine.

All around the district of Anantapur the Collectors were very busy in sending report after report about the misery of people and were actively undertaking relief measures as early as February 1891 in the neighbouring Princely State of Mysore, the Bellary district, the Nizam territory, the Kurnool and Kadapa districts. The Collector of

Anantapur was the last of the region to undertake relief measures very late in December 1891, even though the district was worse than all its neighbouring districts, with the lone exception of Bellary. W.G. Underwood was the Collector of Anantapur during the Khara famine and earlier for eighteen months. His tenure was of proverbial nature. Even though it was a famine year, the Collector, instead of rendering relief measures, resorted to more collections even under Miscellaneous Revenue.

The present attempt, for the first time, show that the 'Dhatu' famine was not for a two year period (1876-77 and 1877-78) as has been depicted so far, but it was for a three year period since it began in 1875-76 itself. The failure to gauge the impending catastrophe in the form of a famine was due to paucity of required information from the Ceded Districts which ultimately led to a successful situation of information collection, registering it and creation of records required on a comparative basis, enabling better management of famine situation, rendering relief to the affected, facilitating the administrators, policy makers and researchers. A comparison is made with the Ryotwari settlement that ravaged from this region during 1800-07 and spread to all the other presidencies in India.

The famine of 1876-78 has been generally called the Dhatu Famine. But the year 1876-77 was Dhatu year and 1877-78 was Eswara year. Of the two years, the famine was very severe during the year 1876-77. The famine of 1876-78 has been the worst one in the history of Indian Famine. The entire South Indian Peninsula suffered very severely. The Madras Presidency, the Mysore State and the Bombay Presidency suffered extremely. In fact, the famine began not in 1876-77, but in the year 1875-76. The earliest symptoms of this impending famine were felt in four districts namely Bellary, Cuddapah, North Arcot and South Arcot.

The rainfall in Ceded Districts for the year 1875-76 was much less than 50% of the rain received during 1874-75. Added to this during 1876- 77, the rainfall alarmingly was a failure. But the rainfall during 1877-78 although half of the quantum of rain. The rainfall during 1878-79 was much improved over that of the three years for

1875-78. The reality was that even this rainfall of 1878-79 was less than the rainfall of 1874-75.

The percentage of rain that fell in the year 1876-77 was as follows: The Presidency rainfall was 75.5%, so this rainfall data clearly reveals that horrible condition were prevailed in Ceded district during the famine which prevailed during 1876-77 the average rainfall of the Ceded districts was 33.08% which as other five Telugu districts of Madras Presidency average rainfall was 74.8% at the same time the average rain fall of other than Telugu speaking districts of Madras. This rainfall data clearly indicate the horrible position of Ceded Districts during the famine of 1876-78.

First Information Report, of the various districts the first alarm of famine had rung from Ceded Districts. The facilities for moving grain by the rail rapidly raised prices everywhere. The activity apparent importation and of railway transit did not indicate any addition to the food-stock of the Presidency, or really afforded security from temporary disaster. So it was specially proved.

The authorities of the Madras Government could not provide the required information. For a second time they sent a telegram on 27th October and asked for sanction for relief works a sum amounting to Rs. 3,77,7700. They also further informed in the telegram that it was impossible to avoid expense or to provide necessary relief from provincial and local funds. But the sanction was not accorded, the production of certain returns which had been asked for, was urged. This is what Sleeman, in his "Rambles and Recollections" complains of the ignorance which prevailed in India in regard to aspects of famines. Therefore, the Governor immediately summoned all Collectors from the districts of Madras and was consulted. Difficulties faced at this meeting ultimately led to the establishment of a Famine Department in December, 1877. This department did not meet the requirements.

System under evolution, consequently the first Famine Commission for Government of India was established which worked during 1878-80 under Sir Richard Strachey. Then came as a result, the Famine Code with a set of guidelines by way of standing preparations with the system of intelligence programmes and estimates; preliminary measures of enquiry and preparation when the rains failed; powers and duties of famine officers wages and allowances; gratuitous relief with it's three kinds; policy regarding closer of relief; other measures of relief, duties of police, medical and sanitary officer in connection with famine and finally preparation of report after famine under thirteen heads.

The origins of the famine from the Ceded Districts along with the Arcot districts, had a propound impact on the collection and registering of information and creation of records resulting in attending to make available relief for the famine stricken people. All these things happened between 1875 and 1879, that is the beginning of the famine to the formulation of Famine Code. Thereafter, the famine that occurred had their respective Famine Reports. With a particular system serving the administrator, famine officer, policy makers, and effective relief apart from the facility, extended to researchers. A similarity may be made in this connection that the Ryotwari settlement emerged in the Ceded Districts during 1800-07 later on extended the other provinces in the country. This was positive one. The present undertaking is a negative one, that the shortage of required information led to useful systematization all over the country.

The Board of Revenue observed on 14 May 1892, that supplies of fodder were scanty throug out. The cattle were emaciated and some of the jungle cattle were reported to be dying. It is revealing to note that the cattle were in miserable condition. Many were dying in Dharmavaram but it is strange to note that numbers were not reported. Similarly no deaths were reported from other taluks. His Highness the Maharaja of Holkar's fund was utilized for purchase of grass. So also, Sir S. Ramaswami Mudaliar's charity. It was reported that great mortality of cattle took place at Kalyandurg without stating the cause. After the recession of famine in August 1892, the Collector's General Report on Anantapur district famine remarked. The famine from which the district has just emerged was one of sorer trial to beasts than to man. It is impossible to say the mortality, but it must have been very great.

It must be observed that all the Collectors of the districts submitted their respective famine reports incorporating the details of cattle mortality, with the exception of Underwood of Anantapur district. For example, Salem, Bellary and Coimbatore districts witnessed a loss of 1,89,561; 1,40,000; 1,27,690 cattle respectively due to famine. In Kadapa and North Arcot, the estimate was over one lakh. Without proper and reliable figures the Famine Commissioner also put the total for Anantapur district at 64,000 without basic information from any of the reports either from the Collector or from the Deputy Collector, Head Assistant Collector, Assistant Collectors, Tahsildars, Deputy Tahsildars and Revenue Inspectors.

The record of the British imperial administration with regard to famine relief received adverse criticisms from many quarters both from within the administration, then and later, and outside the administration. The sufferings of different sections, particularly of the rural areas, knew no bounds. Many a Collector of the distressed districts turned out reasonable work, some known for extraordinary work. Under these conditions, besides imperial interest, many British administrators at the lower level did not work with enthusiasm to alleviate the difficulties of the famine stricken. But, there is no record of any officer of the kind of Underwood of Anantapur who tasted the mild revolt of people of Anantapur through the course of Khara famine as illustrated throughout.

Pratap Kanakalapati

ACKNOWLEDGEMENTS

This region is known for its backwardness and under development about all the successive governments, failed to cater the needs of the common man in society. This area continuously attracting drought and causing distress to the farmers and their dependents. These factors led for chaos and crisis in the social fabric of farming community. It is this region put to inconvenience due to monsoon failure and loss of crops. The Social Scientists suggested that Rayalaseema region needs to be taken special care in providing special assistant and additional support to over come the existing problems. All these factors have been posing hardships to man in this region, in his efforts to achieve self-sufficiency in agriculture production. They created uncertainty in the agriculture output and in turn in the living conditions and living standards of man. Hence, drought proofing works have proved to be of critical significance to agriculture, as is borne out variously by the recordings in its history, the historical factors for the occurrence of repeated drought and famines in Rayalaseema, the factors responsible for the failure of monsoons change of climate conditions and distress of farming community. The impact of famines on the socio, Economic and cultural life of the people of Rayalaseema. The overall development and further initiatives for the betterment of Rayalaseema. The backward region Rayalaseema experienced innumerable droughts and resulted in suffering of farming community and the labourers who depended on agriculture sector the repeated occurrence of famines created havoc in the day to day life of the people of this region. We deeply acknowledge the contribution of academicians, scholars, social scientists, intellectuals, activists and social workers across the country that helped us to take a number of insights from their activities and writings.

Pratap Kanakalapati

Abbreviations

AFPRO	Action for Food program
APRLP	Andhra Pradesh Rural Poverty Reduction Project
BFM	Bengal Famine Mixture
CGWB	Committee for Ground Water Balance
CIDA	Canadian International Development Agency
CV	Coefficient of Variation
DPIP	District Poverty Initiative Project
DWCW	Department of Women & Child Welfare
ESP	Exchangeable Sodium Percentage
FAO	Food & Agriculture Organization
FFW	Food for Work
FIG	Famine Insurance Grant
HRF	Human Rights Forum

ICPRD	Independent Commission for People's Rights and Development
IFC	Indian Famine Codes
INMP	Integrated Nutrient Management Programme
INR	Indian Rupee
LRR	Land Revenue policy Resolution
MPDO	Mandal Parishad Development Officer
NFSM	National Food Security Mission
NMSA	National Mission for Sustainable Agriculture
NWCP	North Western and the Central Provinces,
NWP	North Western Provinces
PET	Potential evapotranspiration
PMRF	Prime Minister's Relief Fund
RADP	Rainfed Area Development Programme
RDRM	Rayalaseema Drought Relief Movement
RRCAPC	Rayalaseema Relief Committee of the Andhra Pradesh Congress
RRLFCBC	Rayalaseema Regional Level Farmers Coalition Building on Combating Drought.
RWS	Rural Water Supply
UNICEF	United Nations Children's Fund

Chapter I

INTRODUCTION

The word "Famine", derived from the Latin word, called, 'Fames', means 'hunger'. Southard in 'Encyclopaedia of social sciences', defines famines as "a state of extreme hunger suffered by population of a region as a result of failure of accustomed food supply". Amartya Sen, in his work, 'Poverty and Famines: Essays on the Entitlement and Deprivation' (1985), defines famines as "a particular virulent manifestation of starvation causing wide-spread death".[1] B.M. Bhatia, in his work, 'Famines in India' (1991), defines, "Famines, under modern conditions, has come to signify an abrupt sharp rise in food prices which renders food beyond the reach of the poor who suffer starvation". He further adds that in a modern famine food may be available at all times in the marker but prices are so high that the poor people cannot purchase it.[2]

A famine is defined as "A famine is a widespread scarcity of food that may apply to any faunal species. This phenomenon is usually accompanied and preceded by regional malnutrition, starvation, epidemic, and increased mortality."[3] Droughts are usually the root cause of famines. In turn droughts where there is a scarcity of life giving water for the crops are usually the direct causes of crop failure in India. The failure of the crops in turn leads to a scarcity of food.

This being said, if India's recent "success" in preventing mass starvation is hardly a momentous achievement, it still remains a creditable one against the background of continuing failures elsewhere. While the "lessons from India" are by no means easy to draw, the rich experience of this country with famine prevention strategies remains well worth scrutinizing.

It is tempting to attribute her relative success in this area to a steady improvement in food production. A close look at the facts, however, quickly reveals the inadequacy of this explanation. Indeed, the period during which the frequency of famines dramatically decreased in India (the first half of this century) was precisely one of the steadily declining food production per head. Since independence in 1947, total food output as admittedly grown at a healthy rate, but per capita food production levels have not increased significantly they appear, in any case, to remain lower than late nineteenth century levels, and also lower than per capita food output levels in many countries affected by famines today. Moreover, the increase of production as resulted first and foremost in the reduction of imports and the accumulation of increasingly large stocks, so that the availability of food as stayed remarkably stagnant over the last 30 years. Last but not least, almost every year large and heavily populated parts of India suffer from devastating droughts which, through the "entitlement failures" they threaten to precipitate, remain quite capable of causing large-scale starvation.

It is more plausible to attribute the disappearance of mass starvation in India during recent decades to the overall evolution of the economy. Sources of livelihood for the rural population are increasingly diversified, and in some areas at least the rapid advance of productivity in agriculture as substantially raised general living standards and further reduced the insecurity of rural life. The Government's general food policy, through far from impervious to criticism has largely succeeded in stabilizing food prices and reducing the correlation between consumption and production. In many states a wide array of more or less successful income support schemes provides a measure of protection against destitution to poor households, and they some accounts at least a discernible trend towards decreasing poverty has emerged since the nightmare of the mind-sixties.

But even this optimist interpretation of recent changes in economic opportunities and policies does not seem to be quite enough to account for the prevention of famines. In the non-irrigated, semi-arid parts of India the stagnation or near-stagnation of yields, population pressure and the increasing frequency of droughts keep the rural population at the mercy of the monsoon, the vulnerability of improvised classes (particularly agricultural labourers) remains an extreme one, and the need persist for a very extensive and expensive relief system. When food prices have assumed unusual proportions (as in 1966-67, 1972-73, 1979-80 and 1985-87), this relief system has been heavily taxed, and has played an undeniable and crucial role in averting large-scale starvation.

Colonialism and Famines

The relationship between colonialism and famines explains why there were frequent occurrences of famines in the latter half of the nineteenth century India, in spite of development of means of communications such as Railway and roads. Here colonial exploitation occupies centre state. Colonial exploitation transformed according to the changing needs of the British economy.[5] Indian economy was given subordinated position in overall colonial economic frame work. Eventually, India made as a "classic colony". India become an exporter of raw materials, food grains and a place to invest capital and became a good market for the machinery made goods from England.[6]

The exploitation of India began in 1757 with the founding of British rule in Bengal. East India Company uses political power to monopolise India's trade. It forced weavers to sell their products at cheaper prices and supplied raw materials at higher price.[7] It made huge profits. By 1813, Indian trade was opened. Free trade was introduced, which was called 'one way traffic'. The machine made goods from England were allows in to the country at nominal duties. Simultaneously, India goods were made to pay higher duties to the entire England and other European countries.[8] Therefore Indian goods were lost both domestic as well as foreign market. Indian indigenous industries were ruined. The industrialization was clamped on people were forced to resort to cultivation in the absence of alternative employment, therefore agriculture was over crowded.[9]

Land revenue settlements were introduced to secure the stable income to meet the expenditure of conquest of India. Poverty rights were introduced, land became a saleable commodity. Land revenue was demanded in cash. So, peasants produced cash crops for the market.[10] Earlier they used to produced only for own consumption and local needs, commercialization of agriculture set in Indian market was integrated into world market.

Frequent assessment of land revenue lead to became one of the highest in the world. When crops were failed, cultivators were forced to lend money from moneylenders for paying the land revenue to the state. Thus peasants were caught in the vicious circle of debt and put to lot of inconvenience.[11]

In this process, cultivators lost their lands to the money lenders of form a class of agriculture labourers who depended on daily wages. People became poor as a whole. Poverty was rampant everywhere, peoples purchasing power was declined.

Colonial rule in India firmly believed in the ideas of classical political economy. It strictly adhered to the notion of free trade between metropolis and colonies. By 1860, India became a classic colony. Exports from India increased constantly over a period of time and were always exceeded imports. They became surplus for which India never get back in return. Food grains were one of the major constituent of theses exports. British rulers hardly showed any interest to intervene free trade in the public interest. They never stop the export of food grains during the famines distress.[12] This attitudes towards trade was worsened the famine conditions.

The poverty of India was not because of lack natural resources or geographical phenomena. It was an end product of colonial exploitation. Therefore the frequent occurrences of famines in later part of the nineteenth century India were manmade.

Famines in Madras Presidency

The poverty of the people was culminated in a series of famines which ravaged all parts of India in the second half of the nineteenth century. There were around 20 famine years between 1860 and 1908. Acceding to one estimate, nearly 29 million people died during famines from 1854-1901.

South India, particularly Madras Presidency also witnessed similar experience of having a series of famines during the period between 1870 and 1900. Four major famines were occurred in 1876-78, 1891-92, 1896-97, and 1899-1900.[13] The immediate cause for the most these famines was the failure of monsoons in bringing rain. So, the failure of crops and increase in the price of food grains. This was coincided by decline in employment and purchasing power. Therefore the famine distress, Governmental relief efforts were either failed or not adequate. The starvation deaths sometimes accompanied by epidemics such as Cholera, Malaria, etc. that is why famine mortality was increased.[14]

The great famine of 1876-1978 was ruined several parts of the presidency including the black spot of ceded districts. It was spared across an area about 83,000 sq miles, and it affected the population of 19.4 millions among which 4 million people died.[15]

The famine of 1891-92 affected many parts. It was spread across an area about 31,069 square miles. It affected a population of 4.98 millions. The later part of famine was witnessed the out-break of Cholera. It increased the famine deaths. The death rate in affected districts was higher than the normal, particularly in ceded districts. For instance, the death rate in Kurnool was 44.1 per mile against normal about 31.8 per mile.[16] The famine of 1896-97 was one of the most disastrous. It was spread across the area about 23,351 square miles and the population affected was 5,048. The famine of 1899-1900 was described by Lord Curzon as a severest and most terrible of all famines that had afflicted country in the nineteenth century, around 25 lakh people died.[17]

In the economic history of agriculture there exist similarities and variations across sub regions, the analysis of which forms important aspect of our study.[18] There is an implicit recognition in many recent studies that in understanding the dynamics of agrarian change (both structural and developmental) in India colonial rule.[19]

The study of agriculture, rural economy, commercialization and the changes in agrarian relations in "dry area" is a fascinating study. As result of commercialization production and productivity increased. There was shift from food crops and non-food crops. The introduction of Railways brought changes in the market conditions. There is simple evidence of land transfer in the twentieth century. Lands were concentrated in the hands of some higher sections of society, which brought changes in the agrarian relations. The poor farmers went into the clutches of money lenders. The real beneficiaries of commoditization were the land lords and money lenders. Admittedly commercialization brought rapid changes in agrarian relations in dry area of Rayalaseema in Andhra.[20]

Famines and drought are two sides of the means same coin "Famine" means "extreme scarcity of food" and "drought" as "continuous dry weather". Drought can be defined or viewed from different angles. Broadly one can divide the dimensions of drought into two — namely (1) Physical, (2) Functional. Meteorological, hydrological and climatic factors that cause drought can be termed as the physical features of drought. Its socio-economic dimension can be termed as the functional feature. Drought is as much a socio-economic phenomenon as it is a meteorological and hydrological phenomenon. In fact drought is a complex feature of all the above said factors.

The various theories on regional variations like "Positive Theory or Martian Historical Approach". To explain this concept point out that it is not only the level of rainfall. But also the social attitudes and institutional factors therein are crucial factors that influence the drought. There can be no second opinion about the view that the ultimate effects of drought percolates to curb the living entities. The biotic content of the region is crippled due to adverse environmental conditions of a drought. Thus the severity of the drought is expressed in the form of biological degradation with social and economic distress.

The ultimate solution that man finds against this natural disaster is migration. Migration to another settlement where he will be able to find employment to seek out a living, generally the migration is by able bodied individuals in search of employment so as to earn and sustain the other members of the family, who normally remain in the native village. In other cases the whole family migrates in search of livelihood. But in any case, the new entrant into an alien society cannot afford to enjoy the social security and freedom.

Unfortunately, during the last several centuries many parts of our country have been prone to severe drought and famine conditions due to several factors namely political instability, internal chaos, spreading of severe infectious diseases like plague, cholera, etc. drastic change in climatic conditions and rapid deforestation.

The impact of geographical and ecological conditions on the History and Culture of any land or region are of great significance. Generally speaking the socio-economic structure, ideology and the nature of a particular.[21] State are determined by the Biophysical and Geo-cultural nature of that region. Recently several scholars (Brook 1973, Wittfogel 1959) recognized the importance of geographical location in the formation of hydraulic societies.[22]

Drought is a state of persistent imbalance between water availability and water need, particularly in the context of agriculture. The Irrigation Commission of India identified the drought prone areas as those areas with a rainfall less than 1000 mm and even 75 percent of this rainfall is not received in 20 percent or more of the years and where irrigation is less than 30 percent of the cropped area?

Thus according to the Irrigation Commission of India's criteria Anantapur District in Andhra Pradesh is a hard core drought area with less than 750 mm rainfall for more than 75 percent of the years and the irrigated areas is less than 25 percent of the cropped area.

The idea was to look at those conditions in terms of *processes.* Too often, poverty and deprivation get covered as *events.* That is, when some disaster strikes, when people die. Yet, poverty is about much more than starvation deaths or near famine conditions. It is the sum fetal of a multiplicity of factors. The weightage of some of these varies from region to region, society to society, culture to culture. But at the core are fairly compact number of factors. They include not just income and calorie intake, land, health, education, literacy, infant mortality rates and life expectancy are also some of them. Debt, assets, irrigation, drinking water, sanitation and jobs count too.

Poor nutrition can impair both mental and physical growth and they can suffer its debilitating impact all their lives. A person lacking minimal access to health at critical moments can face destruction almost as surely as one in hunger. What *is* the access of the hundreds of millions of rural poor to health and education? Do they enjoy the same rights and entitlements as other Indians? If not, what prevents them from doing so? Often, the forms of exploitation that breed and sustain poverty get no more than a cursory glance.

An 'expert group' set up by the Planning Commission submitted its report on the 'Estimation of Proportion and Number of Poor'. The group, which included some of India's leading economists, recommended changes in the Commission's methods of estimating poverty. Their approach found the number of those below the poverty line to be 312 million or close to 39 percent of the population. Now, the government says a later survey than that used by the group shows that those below the official poverty line came down to 19 percent of the population in 1993-94. To get to this result, the Planning Commission has, in part, recycled old, discredited methods of calculation. In the process, it has done away with the suggestions of its own expert group.

The Government of India and the World Summit for Social Development in Copenhagen, presented a document saying 39.9 percent of Indians were below the poverty line. It was, after all, begging for money from donors, the more the money. At home, less than 300 days later, it produced the 19 percent estimate. The people in this work are much like millions and millions of other Indians in many other districts of the poor in India, around 40 percent are landless agricultural labourers. Another 45 percent are small or marginal farmers of the remaining, 7.5 percent are rural artisans, 'others' make up the rest. Most of those in the districts I visited belonged to the first two groups.

Study Area

The Rayalaseema region consists of four districts, i.e. Anantapur, Chittoor, Kadapa and Kurnool. This region is known for its backwardness and under development about all the successive governments to failed to cater the needs of the common man in society. The Four Districts continuously attracting drought and causing distress to the farmers and their dependents, these factors led for chaos and crisis in the social fabric of farming community. It is this region put to inconvenience due to monsoon failure and loss of crops. The Social Scientist suggested that Rayalaseema region needs to be taken special care in providing special assistant and additional support to overcome the existing problems.

All these factors have been posing hardships to man in this region, in his efforts to achieve self-sufficiency in agriculture production. They created uncertainty in the agriculture output and in turn in the living conditions and living standards of man. To alleviate the efforts of these unfavorable forces of nature and also to meet the ever increasing requirements of the ever growing population agriculturist in this region as elsewhere, had to devise artificial waves and means of utilizing water in their agriculture operations with the intentions and aims of increasing agricultural productions and ensuring self-sufficiency that such efforts have been relentlessly done is made out from the study. Hence, drought proofing works have proved to be of critical significance to agriculture, as is borne out variously by the recordings in its history.

Objectives

- To study and analyse, the historical factors for the occurrence of repeated drought and famines in Rayalaseema.
- To analyse the factors responsible for the failure of monsoons change of climate conditions and distress of farming community.
- To analyse the impact of famines on the socio, economic and cultural life of the people of Rayalaseema.
- To evaluate and assess the overall development and further initiatives for the betterment of Rayalaseema.

Methodology

The proposed study aims to employ a combination of historical, descriptive and analytical approach. The present study will look into different parameters and paradigms of famines and agrarian conditions in Andhra Pradesh. The study is both analytical and descriptive. Simple statistical and mapping techniques are employed to facilitate precise explanation. Development strategies to combat drought and studies are based on the primary and secondary data.

Hypothesis of the Study

The Backward Region Rayalaseema experienced innumerable droughts and resulted in suffering of farming community and the labourers who depended on Agriculture sector the repeated occurrence of famines created havoc in the day to day life of the people of this region. After examining the multiple problems faced by the peoples several inferences have been drawn which are pertinent issues to be discussed in detailed for restructuring and to overcome the age old problems that are being faced by the region.

Review of Literature

There are different views on the causes for famines in the modern times: T.R. Malthus in his work *'First Essay on Population'* (1961) views famines are the last resort of nature in ensuring balance between population and resources.

Ester Boserup, in his work, *'The condition of Agriculture Growth'* (1965), differs with Malthus. He argues that population pressure instead of being a cause of demographic catastrophe it became a driving force behind agrarian production.

Mike Davis, 2001, *"Late Victorian Holocausts: El Nino Famines and the Making of the Third World,"* recording the past can be a tricky business for historians. Prophesying the future is even more hazardous. In 1901, shortly before the death of Queen Victoria, the radical writer William Digby looked back to the 1876 Madras famine and confidently asserted, "When the part played by the British Empire in the nineteenth century is regarded by the historian 50 years hence, the unnecessary deaths of millions of Indians would be its principal and most notorious monument." Who now remembers the Madras is?

In Late Victorian Holocausts, Mike Davis charts the unprecedented human suffering caused by a series of extreme climactic conditions in the final quarter of the nineteenth century. Drought and monsoons afflicted much of China, southern Africa, Brazil, Egypt and India. The death tolls were staggering around 12 m Chinese and over 6 m Indians in 1876-1878 alone. The chief culprit, according to Davis, was not the weather, but European empires, with Japan and the US. Their imposition of free-market economics on the colonial world was tantamount to a "cultural genocide".

These are strong words, yet it's hard to disagree with them after reading Davis's harrowing book. Development economists have long argued that drought need not lead to famine, well-stocked inventories and effective distribution can limit the damage. In the nineteenth century, however, drought was treated, particularly by the English in India, as an opportunity for reasserting sovereignty.

William was Lord Lytton, son of the Victorian novelist Edward Bulwer-Lytton (it was a dark and stormy night...) after whom, today, a well-known bad writing prize is named during 1876. Lytton, widely suspected to be insane, ignored all efforts to alleviate the suffering of millions of peasants in the Madras region and concentrated on preparing for Queen Victoria's investiture as Empress of India. The highlight of the celebrations was a week-long feast of lucullan excess at which 68,000 dignitaries heard her promise the nation "happiness, prosperity and welfare".

Lytton believed in free trade. He did nothing to check the huge hikes in grain prices, Economic "modernization" led household and village reserves to be transferred to central depots using recently built railroads. Much was exported to England, where there had been poor harvests. Telegraph technology allowed prices to be centrally coordinated and, inevitably, raised in thousands of small towns. Relief funds were scanty because Lytton was eager to finance military campaigns in Afghanistan. Conditions in emergency camps were so terrible that some peasants preferred to go to jail. A few, starved and senseless, resorted to cannibalism. This was all of little consequence to many English administrators who, as believers in Malthusianism, thought that famine was nature's response to Indian over-breeding.

It used to be that the late nineteenth century was celebrated in every school as the golden period of imperialism. While few of us today would defend empire in moral terms, we have been long encouraged to acknowledge its economic benefits. Yet, as Mike Davis points out, "there was no increase in India's per capita income from 1757 to 1947". In Egypt, too, the financial difficulties caused to peasants by famine encouraged European creditors to override the millennia-old tradition that tenancy was guaranteed for life. What little relief aid reached Brazil, meanwhile, ended up profiting British merchant houses and the reactionary sugar-planter classes.

The European "locusts" did not go unchallenged, rioting became common banditry increased. In China, drought-famine helped to spark the boxer uprising. In Europe, the *fin de siècle* was largely an opportunity for pale-faced men to wear purple cummerbunds and spout rotten symbolist poetry; for colonized peoples it genuinely seemed to presage mass extinction. It was, says Mike Davis, "a new dark age of colonial war, indentured labour, concentration camps, genocide, forced migration, famine and disease."

Mike Davis's attention to the importance of environment may recall the work of the annales school of historians, but he is far more radical than any of them. His writing, both here and in such classic books as City of Quartz and Ecology of Fear, is closer to that of Latin American intellectuals such as Ariel Dorfman and the Uruguay, Eduardo Galeano, who for decades have spotlighted capitalism's casual abuse of the third world and who have sought to champion the poor and dispossessed. Such commitment, forcefully and lucidly expressed, is unfashionable these days.

"Class" may be passé in academic circles, yet the catalogue of cruelty Davis has unearthed is jaw-dropping. A friend to whom I lent the book was reduced to tears by it. Late Victorian Holocausts is as ugly as it is compelling. But, as Conrad's Marlow said in Heart of Darkness: "The conquest of the earth, which means the taking away from those who have a different complexion and slightly flatter noses than ourselves, is not a pretty thing when you look at it too much."

Kallie Szczepanski, *The 1899-1900 Famine in India* in 1899, the monsoon rains failed in central India. Drought parched crops over an area of at least 1,230,000 square kilometers (4,74,906 square miles), impacting almost 60 million people. Food crops and livestock died as the drought stretched into a second year, and soon people began to starve. The Indian Famine of 1899-1900 killed millions of people, perhaps as many as 9 million in all.

Many of the famine victims lived in British-administered sections of colonial India. The British Viceroy of India, Lord George Curzon,

Baron of Kedleston, was concerned with his budget and feared that aid to the starving would cause them to become dependent on hand-outs, so British aid was seriously inadequate, at best. Despite the fact that Great Britain had been profiting greatly from its holdings in India for more than a century, the British stood aside and allowed millions of people in the British Raj to starve to death. This event was one of several that inspired calls for Indian independence, calls that would increase in volume over the first half of the twentieth century.

B.M. Bhatia, in his work, 'Famines in India, 1860-1900' (1991), attributes to the British economic policies as a major cause for the frequent occurrence of famine in the late nineteenth century India. He says that the introductions of capitalistic features in agriculture, ruination of indigenous industries were resulted in the poverty of the people.

Nimmakayala Venkata Subba Reddy, 2010, *Famine in Rayalaseema, 1876-78*, in the later half-of the nineteenth century, South India witnessed a series of severe famines, of them one during 1876-78 was the most severe and devastating. The whole of Madras Presidency, the Bombay Presidency and the princely state of Mysore were affected 50 lakhs of people and perished on account of this famine. The relief offered by the British government was terribly in sufficient. The rain shadow region of the then ceded districts, the present Rayalaseema districts namely Bellary (inclusive of Anantapur District), Cuddapah and Kurnool Districts suffered the worst. Nearly 7 lakhs of people were died in this region. In this book thoroughly discussed aspects like scanty rainfall, loss of crops, decreasing cultivation, shortage of food grains, increasing prices, etc. It is an attempt micro level research of the famine of 1876-78 in the Rayalaseema region.

B.M. Bhatia, *Famines in India: A Study in Some Aspects of the Economic History of India, 1860-1945*, (1963) Famines Food supply India.

B.M. Bhatia, *Food Security in South Asia*, 1985 - Agriculture.

Stephen Devereux, *The New Famines – Why Famines persist in an Era of Globalization,* 2007. Copyright, Exploring the paradox that is the persistence of famine in the contemporary world, this book looks at the way the nature of famine is changing in the face of globalization and shifting geo-political forces.

J.W. Parker and Son,1883, *The Saturday Review of Politics, Literature, Science and Art*, Volume 50

W. Simpkin and R. Marshall, 1792, The Critical Review, or, Annals of Literature. They, had been shut up in their fortress, besieged by the Indians, and driven by *famine* to the utmost distress, obliged to cat serpents, West-India islands, which shows its powers by tumours on the skin, is infectious and cured.

H.K. Mishra, *Famines and Poverty in India, 1991,* Although much has been written on the socio-political and religious history of Orissa during the period under study, little or no attempt has so far been made to study the economic condition of Orissa and much less the problems of poverty and famines in the said period.

Ghanshyam Shah, *Social Movements in India: A Review of Literature*, 2004, social movements hasn't been a popular topic with researchers, making up less than 3 percent of all studies in history, political science, sociology and anthropology sponsored by the Indian Council of Social Science Research (ICSSR) up to the mid-nineties. The research has had an 'institutional' or 'government' skew, in that, the study of the politics of the masses has been largely ignored. There are reasons of history behind this, but what has been consistently lost sight of is the fact that in the absence of an understanding of the politics of the masses, the functioning of the state can be understood only partially. This volume is a revised and enlarged edition of the author's review of literature on social movements in India, first commissioned by the ICSSR. After careful deliberation on the 'ideal' definition of a 'social movement', the author adopts for this volume the loose idea of 'non-institutionalised collective political action striving for social and political change'. On the basis of the socio-economic characteristics of participants and the issues involved, this volume makes a nine-fold classification of social movements, peasant movements, tribal movements, dalit movements, backward caste movements, women's movements, working class movements, student's movements, middle class movements, human rights and environmental movements (added in this edition). This book is important as much for filling a scholarly lacuna in social science studies as for proposing and executing an orderly classification of literature on social movements in modern India. The original, shorter, monograph received an enthusiastic response from both scholars and laypersons, and this volume is likely to be welcomed similarly.

Patrick O'Sullivan, *The Meaning of the Famine*, 1997, we now need to move swiftly forward, assuming that reader knows something about events in Ireland suffered its great famine. In 1845-50, India continued to suffer its famines. In 1857, there occurred the events that are referred to in British textbook as the Indian mutiny, or sometimes the Sepoy Mutiny and in Indian histories as the.

David Arnold: in his article, *'Famines in peasant consciousness and peasant action: Madras 1876-78'* (1984) views that the working population who depends on wages, particularly agricultural labourers and artisans are most vulnerable during the famine crisis.

In the latter half of the nineteenth century India witnessed the increase in the frequently and intensify of famines and perceptions are differ on this, National Economists such as Dadabhai Naoroji, D.E. Wacha, G.V. Joshi, K.T. Telang, M.G. Ranade, G.K. Gohale, G.S. Iyer are opined that famines are not always due to natural calamities. But due to the defective and short sighted policies of colonial rule. So, famines also are manmade. Colonial view differs with nationalist opinion. It attributes the causes to natural calamities, over population and extravagant social habits and customs of the Indian people. India population reached its saturation point. Therefore famines were to bring equilibrium between population and resources. So nothing could be done to prevent famine death.

Ramesh Chandra Dutt in his work, *'Famines and land Assessment* ' 1985, argues that the famines are due to the chronic poverty of the cultivators because of over assessment of land revenue. He feels that permanent land settlement was the only remedy. Hari Sanker Srivastava in his work, *'History of India Famines and Development of Famine Policy, 1815-1918'*, (1991), opinions that famines are due to colonial political economic and social policies introduces in India.

Ira Klein in his article, *'when rains failed: famine, relief and mortality in British India'* (1984), opines that modernisation of India was the main cause. He explains that modernisation process in India created two sections of people. One section consisted of land lords, middlemen and money lenders who benefited from the famine distress. The other section of people consisted of small peasants, labourers who suffered from famine distress.

Thomas Keneally, Three Famines: Starvation and Politics, 2010, Through the lens of three of the most devastating food crises in modern history, the Gorta Mor of British-ruled Ireland, the great famine of British-ruled Bengal in 1943, and the string of famines that plagued Ethiopia during the 1970s and 1980s, Booker Prize-winning author Thomas Keneally vividly evokes the terrible cost of mass starvation at the level of the individual who starves and the nation that watches. Famine is widely misunderstood as a completely natural catastrophe. Keneally recounts that while the triggers-crop, pestilence, and drought-are natural, the political and ideological choices that prolong famine are man-made. Government neglect and individual venality, not food shortages, are historically the causes of sustained, widespread hunger. In Ireland, British authorities ignored the existence of a food crisis while the famished fed on diseased cattle and human remains. In Bengal, where over four million starved to death, Field Marshal Archibald Wavell's reports of people dying in Calcutta's streets and demands for relief resulted in little more than a mocking cable from Winston Churchill asking, why, if food was so scarce, hadn't Gandhi died yet? In Ethiopia, Mengistu Haile Mariam arranged for 4,00,000 bottles of whisky to ship to Addis Ababa from Britain to celebrate the tenth anniversary of the revolution that put him in power, while one person died every twenty minutes in Korem. These three famines are stark examples of how throughout history, racial preconceptions, administrative neglect, and incompetence have been more lethal than the initiating blights or crop failures. Keneally's startling narrative history is a sobering warning to the authorities in charge of mercy in our time to stop making choices that feed famine instead of the starving.

N. Jayapalan, *Economic History of India*, 2008 for a proper understanding of India s economy as we find today with its fast development, a study of its past economic conditions is necessary. India s economic history spans nearly 5000 years and dates back to Indus Valley Civilization the two prominent cities of which Mohenjodaro and Harappa were big industrial centres having trade relations with West Asia and other parts of the world. Several sources provide us valuable information about the economic condition of the people of India at various stages of history from Vedic literature we get a detailed description of the economic life of the Aryans. The Buddhist literature, particularly *Jatakas* and *Tripitaka* give us a glimpse of the economic condition of India from 650-321 BCE when every village was a self-sufficient unit. Kautilyas *Arthashastra*, *Megasthenes Indica* and Vishakhadatta's *Mudrarakshasa* give detailed description of the economic systems of the Mauryan period. The accounts of Chinese travellers Fa-hien and It-sing are remarkable evidences of the socio-economic life in Gupta period. Historians like Manucci tell us about the economic reforms undertaken during Mughal rule in India. The British who carved out their empire in India in the seventeenth century, drained India of its wealth, destroyed the self-sufficient character of villages, increased agricultural indebtedness, and gave rise to a capitalist class in India with the growth of new urban centres. After the achievement of independence in 1947 started the process of nation-building through economic planning with the liberalization of Indian economy in the 1990s, the country has been put on a high growth path and is making fast economic progress. The book captures the salient features of India s economic history in chronological order. It will fulfill the needs of students and teachers of this subject and prove immensely useful to the aspirants of Civil Services and other competitive examinations.

Atreyi Biswas, 2000, *Famines in Ancient India: A Study of Agro-Economy from Prehistoric to Early Historic Period,* A novel study of a very complicated subject, the author has analyzed the archaeological data in ecological contexts. The volume is based on an extensive study of archaeological reports. This book is an essential reading for scholars and experts dealing in the concerned area.

B.R. Tomlinson, *The Economy of Modern India, 1860-1970, 1996,* This is the first comprehensive and interpretative account of the history of economic growth and change in colonial and post-colonial India.

Dr. Tomlinson draws together and expands on the specialist literature dealing with imperialism, development and underdevelopment, the historical processes of change in agriculture, trade and manufacture, and the relations among business, the economy and the state. What emerges is a picture of an economy in which some output growth and technical change occurred both before and after 1947, but in which a broadly based process of development has been constrained by structural and market imperfections. Tomlinson argues that India has thus had an underdeveloped economy, with weak market structures and underdeveloped institutions, which has since 1860 profoundly influenced the social, political and ecological history of South Asia.

Radhey Shyam Chaurasia, 2002, *History of Modern India, 1707-2000 A.D.*, tremendous progress has been made in India during the modern period. British rule unified India, gave new ideals of parliamentary government and established factories, railways, telephone, etc. due to development of new scientific weapons and impact of industrial revolution, East India company was able to defeat Indian powers and succeeded in establishing British rule in India, Burma and Ceylon. In 1857, great rebellion took place which ended rule of east India company and British parliament in the name of queen and king began to rule all over India through the secretary of state for India and the viceroy of India. The book is divided into two parts, first part deals with Anglo-French wars, Maratha and Sikh wars and wars with other small powers and role of different Governor Generals such as Clive, Warren Hastings, Cornwallis, Wellesley, Hastings and Dalhousie, etc. The socio-religious movements took place during this period and Brahmo Samaj was established by Raja Ram Mohan Roy, Arya Samaj by Swami Dayanand, Ram Krishna Mission by Swami Vivekananda. The Reform movements of Muslims and Sikhs also took place. The second part deals with the constitutional developments and nationalist movement and the role played by the eminent leaders during this period. Ultimately, India became free on 15th August, 1947 and constitutional parliamentary government was established and India became the largest democracy of the World. India was divided; Pakistan came into existence, which gave rise to conflicts between these two powers. Though in 1971,

Pakistan was divided and Bangladesh came into existence but conflict is still continuing. After Independence, India has made great progress and she is now one of the mightiest powers on earth with nuclear weapons and viable economy. From 1947-2002, tremendous progress has been made in scientific inventions, art, literature and in other social aspects which have been described in brief, unfortunately, modern history of India has been written by British writers with imperialist point of view. In this book an attempt has been made to give objective outlook.

Frank Dikötter, 2010, *Mao's Great Famine: The History of China's Most Devastating Catastrophe, 1958-1962,* "Between 1958 and 1962, China descended into hell. Mao Zedong threw his country into a frenzy with the Great Leap Forward, an attempt to catch up to and overtake Britain in less than 15 years The experiment ended in the greatest catastrophe the country had ever known, destroying tens of millions of lives." So opens Frank Dikotter's riveting, magnificently detailed chronicle of an era in Chinese history much speculated about but never before fully documented because access to Communist Party archives has long been restricted to all but the most trusted historians. A new archive law has opened up thousands of central and provincial documents that "fundamentally change the way one can study the Maoist era". Dikotter makes clear, as nobody has before, that far from being the program that would lift the country among the world's superpowers and prove the power of Communism, as Mao imagined, the Great Leap Forward transformed the country in the other direction. It became the site not only of "one of the most deadly mass killings of human history", at least 45 million people were worked, starved, or beaten to death, but also of "the greatest demolition of real estate in human history", as up to one-third of all housing was turned into rubble. The experiment was a catastrophe for the natural world as well, as the land was savaged in the maniacal pursuit of steel and other industrial accomplishments. In a powerful mesghing of exhaustive research in Chinese archives and narrative drive, Dikotter for the first time links up what happened in the corridors of power-the vicious backstabbing and bullying tactics that took place among party leaders-with the everyday experiences of ordinary people, giving voice to the dead and disenfranchised. His magisterial account recasts the history of the People's Republic of China.

Tim Pat Coogan, 2012, *The Famine Plot: England's Role in Ireland's Greatest Tragedy,* During a Biblical seven years in the middle of the nineteenth century, Ireland experienced the worst disaster a nation could suffer. Fully a quarter of its citizens either perished from starvation or emigrated, with so many dying en route that it was said, "you can walk dry shod to America on their bodies." In this grand, sweeping narrative, Ireland's best-known historian, Tim Pat Coogan, gives a fresh and comprehensive account of one of the darkest chapters in world history, arguing that Britain was in large part responsible for the extent of the national tragedy, and in fact engineered the food shortage in one of the earliest cases of ethnic cleansing. So strong was anti-Irish sentiment in the mainland that the English parliament referred to the famine as "God's lesson". Drawing on recently uncovered sources, and with the sharp eye of a seasoned historian, Coogan delivers fresh insights into the famine's causes, recounts its unspeakable events, and delves into the legacy of the "famine mentality" that followed immigrants across the Atlantic to the shores of the United States and had lasting effects on the population left behind. This is a broad, magisterial history of a tragedy that shook the nineteenth century and still impacts the worldwide Irish Diaspora of nearly 80 million people today.

Brian M. Fagan, Floods, Famines, and Emperors: El Ni*n*o and the Fate of Civilizations, 2009, In 1999, few people had thought to examine the effects of climate on civilization. Now, due in part to the groundbreaking work of archaeologist Brian Fagan, climate change is a central issue. Revised and updated ten years after its first publication, Floods, Famines and Emperors remains the definitive account of how the world's best-known climate event had an indelible impact on history.

Narendra Nath Kher, 1973, *Agrarian and Fiscal Economy in the Mauryan and Post Mauryan Age (cir. 324 B.C.-320 A.D.).*

P. Sainath, 2000, Everybody loves a good drought, The human face of poverty The poor in India are, too often, reduced to statistics. In the dry language of development reports and economic projections, the true misery of the 312 million who live below the poverty line, or the 26 million displaced by various projects, or the 13 million who suffer from tuberculosis gets overlooked. In this thoroughly researched study of the poorest of the poor, we get to see how they manage, what sustains them, and the efforts, often ludicrous, to do something for them. The people who figure in this book typify the lives and aspirations of a large section of Indian society, and their stories present us with the true face of development.

David Ludden, 1999, *An Agrarian History of South Asia*, Part 4, Volume 4, David Ludden provides a comprehensive historical framework for the understanding of regional diversity of agrarian South Asia. Adopting a long-term view, he treats South Asia not as a single civilization territory, but as a patchwork of agrarian regions, with their own social, cultural and political histories. He traces these histories from medieval times to the present. As a comparative synthesis of the literature on agrarian regimes in South Asia, this will be a valuable resource for students of agrarian and regional history, as well as comparative world history.

Suzanne Franks, Reporting Disasters: Famine, Aid, Politics and the Media, 2013, The media reporting of the Ethiopian Famine in 1984-5 was an iconic news event. It is widely believed to have had an unprecedented impact, challenging perceptions of Africa and mobilising public opinion and philanthropic action in a dramatic new way. The contemporary international configuration of aid, media pressure, and official policy is still directly affected and sometimes distorted by what was as this narrative shows also an inaccurate and misleading story. In popular memory, the reporting of the Ethiopian famine and the resulting humanitarian intervention were a great success. Yet alternative interpretations give a radically different picture of misleading journalism and an aid effort which did more harm than good. Using privileged access to BBC and Government archives, Reporting Disasters examines and reveals the internal factors which drove BBC news and offers a rare case study of how the media can affect public opinion and policymaking. It constructs the process that accounts for the immensity of the news event, following the response at the heart of government to the pressure of public opinion.

And it shows that, while the reporting and the altruistic festival that it produced triggered remarkable and identifiable changes, the ongoing impact was not what the conventional account claims it to have been.

Amartya Sen (1983) *Poverty and Famines: An Essay on Entitlement and Deprivation,* Commodities and Capabilities presents a set of inter-related theses concerning the foundations of welfare economics, and in particular about the assessment of personal well-being and advantage. The argument presented focuses on the capability to function, i.e. what a person can do or can be, questioning in the process the more standard emphasis on opulence or on utility. In fact, a person's motivation behind choice is treated here as a parametric variable which may or may not coincide with the pursuit of self-interest. Given the large number of practical problems arising from the roles and limitations of different concepts of interest and the judgement of advantage and well-being, this scholarly investigation is both of theoretical interest and practical import.

Margaret Kelleher (1997) *The Feminization of Famine: Expressions of the Inexpressible*?, Contemporary depictions of famine and disaster are dominated by female images. *The Feminization of Famine* examines these representations, exploring, in particular, the literature arising from the Irish "Great Famine" of the 1840s and the Bengali famine of the 1940s. Kelleher illuminates recurring motifs the prevalence of mother and child images, the scrutiny of women's starved bodies, and the reliance on the female figure to express the largely "inexpressible" reality of famine. Questioning what gives these particularly feminine images their affective power and analyzing the responses they generate, this historical critique reveals striking parallels between these two "great" famines and current representations of similar natural disasters and catastrophes.

Kelleher begins with a critical reading of the novels and short stories written about the Irish famine over the last 150 years, from the novels of William Carleton and Anthony Trollope to the writings of Liam O'Flaherty and John Banville. She then moves on to unveil a lesser-known body of literature-works written by women. This literature is read in the context of a rich variety of other sources, including eye-witness accounts, memoirs, journalistic accounts, and famine historiography. Concluding with a reading of the twentieth-century accounts of the famine in Bengal, this book reveals how gendered representations have played a crucial role in defining notions of famine.

Romesh Chander Dutt (1950) *The Economic History of India Under Early British Rule: From the Rise of the* British Power in 1757 to the Accession of Queen Victoria in 1837, First published in 2000. Routledge is an imprint of Taylor & Francis, an informa company.

S. N. Kulkarni (1990) *Famines, Droughts, and Scarcities in India: Relief Measures and Policies, Onwards and Famine Policy in India with special reference to Maharashtra*. Part 2 of the work is a report on drought and scarcity conditions in Sinnar Taluk during the period 1970-73 and 1983-86 based on detailed investigation of sample households from 11 samples villages.

Jean Dreze and Amartya Kumar Sen (1989) *Hunger and Public Action,* examining the problem of hunger in the modern world and the role public opinion might play in combating it, Dreze and Sen here provide a coherent perspective on the complex nutritional, economic, social, and political issues involved in the analysis of hunger. They explore famine prevention through a series of case studies in Africa and elsewhere, and discuss the problem of chronic undernourishment. Sen was awarded the second Agnelli Prize for the Ethical Dimension in Advanced Societies in March 1990 in recognition of his outstanding contribution to the understanding of the ethical dimension in modern society.

Geoffrey Bussell Masefield (1963) *Famine: Its Prevention and Relief,* these are usually designed with the secondary object of benefiting the country generally and especially of reducing the chances of future famine. In India, irrigation and raod making have been the most favoured subjects.

Paul Robert Greenough, 1982, *Prosperity and Misery in Modern Bengal: the famine of 1943-1944.* This book is a provocative and moving account of the reasons for and the consequences of the Bengal Famine of 1943 to 1944.

Brendan O Cathaoir, *Famine diary*, 1999, A unique record of the Great Famine, written with insight, detachment and empathy. Based on a wide selection of sources - contemporary newspapers, official correspondence and diaries - it provides a graphic picture of conditions in the Irish countryside as the crisis developed. It combines analysis and an overview with a focus on the worst hit areas. The relief efforts are presented where possible, with the help of priests' letters, from the perspective of the poor.

Dessalegn Rahmato, 1991, *Famine and Survival Strategies: A Case Study from Northeast Ethiopia*, What do peasants do in the face of severe food crisis and ecological stress, and how do they manage to survive on their own? This study revolves around a case study conducted by the author in the Awraja (district) in the Ambassel Wollo province in north eastern Ethiopia. This is in the region that was hit hardest by the 1984-85 famine, which Rahmato calls "the worst tragedy rural Ethiopia had ever experienced". The author also critically examines other literature on famine response. The focus of this study is on what happens before famine comes, and how the peasants prepare for it. From a wealth of evidence, the author concludes that the seeds of famine are sown during the years of recovery.

Stephen Devereux, 1994, *Theories of Famine,* Reviews the major theories and critically examines alternative explanations for famine, describing their implications for preventative policies and corrective interventions.

Nibedita Shankar Ray-Bennett, *Indian Famine (1896-1902)*, the famines of 1896-97 and 1899-1900 in India affected almost the entire subcontinent, causing severe distress, debility, and mortality. Unlike the 1876-78 famine, which was largely localized, these famines were widespread and occurred due to an unprecedented shortage of food grains and escalating prices. At that time, these famines were officially described as the most disastrous famines of the nineteenth century due to their extent and severity. The famine of 1896-97 affected Bihar, Bengal, Bombay Deccan, the Deccan districts of the Madras Presidency, the Northwest Provinces and Oudh, the Central Provinces, the Punjab, and Burma. It was estimated the famine tract covered an area of 2,24,828 sq. miles and affected a population of

62.4 million. The highest numbers of deaths (745,376) was recorded in Bombay, followed by 1,87,686 in the Central Provinces, excluding zamindaris (landholders); 1,27,409 in the Punjab and 125,926 in Berar. The famine of 1899-1900 followed in.

Amrita Rangasami in critic on Amartya Sen, *'Failure of Exchange Entitlements Theory of Famines: A Response',* (1985) argues that the entitlements theory only considered with the final phase of famine and it does not talk about advance of famine. He opines that, famine is a process it accompanies by political, economic and social forces that mark the onset of the famine.

Organization of the Book

The main focus of this work is on the Famine and the agrarian system in south India in general and famines in Rayalaseema region in particular. The book has been divided in six chapters. First chapter deals with the introduction, it includes the area of the study, the main objectives of the book. The second chapter discusses the origin, and historical background of Famines, in which a detailed historical evolution has been discussed at length. The third chapter explains the economic deprivation of agrarians. This chapter throws light on the incontinences caused to the agrarians in their day to day life due to severe famines occurred in this region.

The fourth chapter probes into the impact of famines on agrarians conditions in Rayalaseema. The policies of the government to combat the situation and the reactions of the civil society has been discussed by means of talking different sample studies in the study area. The fifth chapter focused on various facets and causes of famine and its impact on agrarian system of drought prone area, to which the Rayalaseema region in general, Ananthapur district in particular fell as a victim right from the historical period to the present day.

The sixth chapter is a conclusive chapter presenting different paradigms of famine issues, social relevance of victims, various view points and observations in the context of famines in Rayalaseema region. In this chapter, a detailed account on the recommendations, reports and the reaction of Civil Society was meticulously enumerated to justify the findings and arguments of the researcher.

Bibliography

1. Amartya Sen, *Poverty and Famines: Essays on the Entitlement and Deprivation*, 1983.
2. B.M. Bhatia, *Famines in India*, 1991.
3. A.C. Sahu, *Diversion of arable lands from food crops to exportable cash crops: One of the causes of frequent famines in India between 1860-1900*, IHC proceedings, modern India section, 41st session Bombay, 1980.
4. A.C. Sahu, *Export of food grains: a potent cause of famine in India between 1860-1900*, IHC proceedings, Modern India section, 39th session Hyderabad, 1978, vol. II.
5. Sugatha Bose (ed.), *Credit Markets and the Agrarian Economy of Colonial India*, OUP, Delhi, 1994.
6. David Hall Mathews, 'Colonial Ideologies of the Market and Famine Policy in Ahmednagar district, Bombay Presidency 1870-84', *Indian Economic and Social History Review*, vol. xxxvi, no. 3, July-September, 1999.
7. Dietmar Rothermund, *Economic History of India from pre-colonial times to 1986*, Manohar Publications, New Delhi, 1988.
8. Irfan Habib, 'Studying a Colonial Economy without Perceiving Colonialism', *Social Scientist*, vol. 12, no. 12, December, 1984.
9. Sabyasachi Bhattacharya and *et al.*, *The South Indian Economy: Agrarian Change, Industrial Structure and State Policy 1914-1947*, OUP, Delhi, 1991.
10. David Washbrook, 'The Commercialization of Agriculture in Colonial India: Production, Subsistence and Reproduction in the Dry South, c. 1870-1930', *Modern Asian Studies*, vol. 28, no. 1, 1994.
11. G. Suseela Raju, 'Critical Analysis on the Impact of Moneylenders on Peasants in Andhra Pradesh (1900-1947)', *ITIHAS*, vol. XVI, no. 2, July-Dec., 1990.
12. V. Ratna Reddy, *Irrigation in Colonial India – A Study of Madras Presidency during 1860-1900*, 1990.

13. David Hall Mathews, *op. cit.*, 1999.

14. David Arnold, 'Famines in peasant consciousness and peasant action: Madras 1876-78', In Ranjith Guha (ed.), *Subaltern Studies III: Writings on South Asian History and Society*, Delhi, 1984.

15. William Digby, *The Famine Campaign in Southern India: Madras and Bombay Presidencies and Province of Mysore, 1876-1878*, Volume 1, 1878, p. 105.

16. Roland Lardinois, *Famine Epidemics and Mortality in South India – A Reappraisal of the Demographic Crisis of 1876-78*, vol. 11, 1985.

17. Neil Charlesworth, *British Rule and the Indian Economy 1800-1984*, Macmillan, London, 1982.

18. Dharma Kumar, *Land and Caste in South India*, Manohar Publishers, New Delhi, 1978.

19. M.S.S. Pandian, *The Political Economy of Agrarian Change: Nanchilnadu 1880-1939*, Sage Publications, New Delhi, 1990.

20. A. Satyanarayana, *Andhra Peasants under British Rule: Agrarian Relations and Rural Economy 1900-1940*, Manohar Publishers, New Delhi, 1990.

21. D.A. Wash Brook, 'Economic Development and Social Stratification in Rural Madras: The Dry Region, 1878-1929', In G. Dewey and A.G. Hopkins (eds.), *The Imperial Impact: Studies in the Economic History of Africa and India*, The Athlene Press, 1978. A. Karl Wittfogel, *Oriental Despotism: A Comparative Study of Total Power*, 1957.

Chapter II

Origin and Historical Background of Famines

Life is a combination of pleasure and pains. Pleasure is a concept, which need not be explained. But the idea of pain needs some attention, any disaster may be at the root of 'Pain'. Among the many standpoints for disaster, 'famine' is the most dreadful one. It is a 'menace to mankind'. That is why in the olden days famine and controlling of famine were the most tumultuous problems to the State. With a changeover of the conditions of the society over a period, and with stringent efforts to erase the dreary of famine, famine-creating conditions appear to have eroded. But it has not been completely extinguished. And in fact with new dimensions and with new phases its prevalence is being experienced. Famine situation of Anantapur district can be given as one example. In this regard it is worthy to assess as to what the so-called concept of famine means and what are its multifarious ways.

Famine is a disaster that occurs as the result primarily of drought, but it can also follow pestilence, windstorm, and human-induced catastrophes such as war and civil strife. Starvation is the result of food shortage. Famine and food emergencies are common events particularly in Africa and Asia, despite many intense and continuing efforts to address the problems. Drought related famines appear to occur with cyclical frequency in many parts of Africa, both along the edges of deserts and in certain rain forest regions.

Famines rarely occur unexpectedly. Food stocks are not often depleted or destroyed suddenly and simultaneously in large communities. Crops in the fields may be destroyed quickly by pests such as locusts, but grain stocks may be rendered totally unfit for human consumption only as a result of moisture damage, infestation or contamination. Non-grain commodities are more vulnerable to damage and loss, especially those that cannot be stored without modern technology such as refrigeration.

More frequently, famine is predictable, in other words, the creeping onset of crop failure or food emergency is predictable from a series of meteorologic, agricultural, political and economic indicators that may be monitored continuously. This process of famine preparedness through surveillance is a major function of the Food & Agriculture Organization (FAO) headquartered in Rome. Where international concern is not hampered by political constraints, the early warning system has begun to work relatively well in recent years, and international appeals for food assistance have been mounted successfully on many occasions. However, the bureaucracy is somewhat cumbersome, and bulk shipments by sea and by land are subject to many delays, so that international response to famine has often been a slow process. Thus, it is extremely important that indicators of an impending famine be closely monitored and measures taken quickly if it appears that a food shortage is developing priority be given to developing strong food and agricultural systems that provide an adequate fall-back resource in threatened areas.

Famines continue to raise deep questions about the performance of economic and political institutions. Famines in British India were severe enough to had a substantial impact on the long term population growth of the century in the nineteenth and early twentieth centuries.

Indian agriculture is heavily dependent on climate a favourable southwest summer monsoon is critical in securing water for irrigating crops. Droughts, combined with policy failures, have periodically led to major Indian famines, including the Bengal famine of 1770,[1] the Chalisa famine, the Doji Bara famine, the Great Famine of 1876-78, and the Bengal famine of 1943.[2] Some commentators have identified British government inaction as contributing factors to the severity of famines during the time India was under British rule. The 1883 Indian Famine Codes, transportation improvements, and changes following independence have been identified as furthering famine relief. In India, traditionally, agricultural labourers and rural artisans have been the primary victims of famines. In the worst famines, cultivators have also been susceptible.[3]

One of the earliest treatises on famine relief goes back more than 2000 years. This treatise is commonly attributed to Kautilya, who recommended that a good king should build new forts and water-works and share his provisions with the people, or entrust the country to another king.[4] Historically, Indian rulers have employed several methods of famine relief. Some of these were direct, such as initiating free distribution of food grains and throwing open grain stores and kitchens to the people. Other measures were monetary policies such as remission of revenue, remission of taxes, increase of pay to soldiers, and payment of advances. Yet other measures included construction of public works, canals, embankments and sinking wells.

The late eighteenth and nineteenth centuries saw increase in the incidence of severe famine. These famines in British India were bad enough to have a remarkable impact on the long term population growth of the country, especially in the half century between 1871-1921.[5] The first, the Bengal famine of 1770, is estimated to have taken the lives of nearly one-third of the population of the region, about 10 million people.[6] The impact of the famine caused East India Company revenues from Bengal to decline to £174,300 in 1770-71. The stock price of the East India Company fell sharply as a result. The company was forced to obtain a loan of £1 million from the Bank of England to fund the annual military budget of between £60,000-1 million.[7] Attempts were later made to show that net revenue was unaffected by the famine, but this was possible only because the collection had been "violently kept up to its former standard".[8] The 1901 Famine Commission found that twelve famines and four "severe scarcities" took place between 1765 and 1858.[9]

Researcher Brian Murton states that the famines recorded after the arrival of the English, but before the establishment of the Indian Famine Codes of the 1880s, bear a cultural bias regarding the stated causes of the famine because they "reflect the view of a handful of Englishmen."[10] These sources, however, contain accurate recordings of weather and crop conditions. Florence Nightingale made efforts to educate British subjects about India's famines through a series of publications in the 1870s and beyond.[11] Evidence suggests that there may have been large famines in south India every forty years in pre-colonial India, and that the frequency might have been higher after the twelfth century. These famines still did not approach the incidence of famines of the eighteenth and nineteenth centuries under British rule.

Causes of Famine

The famines were a product both of uneven rainfall and British economic and administrative policies.[12] Colonial polices implicated include rack-renting, levies for war, free trade policies, the expansion of export agriculture, and neglect of agricultural investment.[13] Indian exports of opium, rice, wheat, indigo, jute, and cotton were a key component of the economy of the British empire, generating vital foreign currency, primarily from China, and stabilising low prices in the British grain market.[14] Export crops displaced millions of acres that could have been used for domestic subsistence, and increased the vulnerability of Indians to food crises. Others dispute that exports were a major cause of the famine, pointing out that trade did have a stabilising influence on India's food consumption, albeit a small one.[15]

The famine of 1876-78, also known as the Great Famine of 1876-78, caused a large migration of agricultural labourers and artisans from southern India to British tropical colonies, where they worked as indentured labourers on plantations. The large death toll, about 10.3 million, offset the usual population growth in the Bombay and Madras Presidencies between the first and second censuses of British India in 1871 and 1881 respectively.[16]

The large-scale loss of life due to the series of famines between 1860 and 1877 was the cause of political controversy and discussion which led to the formation of the Indian Famine Commission. This commission would later come up with a draft version of the Indian Famine Code.[17] It was the Great Famine of 1876-78, however, that was the direct cause of investigations and the beginning of a process that led to the establishment of the Indian Famine code.[18] The next major famine was the Indian famine of 1896-97. Although this famine was preceded by a drought in the Madras Presidency, it was made more acute by the government's policy of *laissez faire* in the trade of grain.[19] For example, two of the worst famine-afflicted areas in the Madras Presidency, the districts of Ganjam and Vizagapatam, continued to export grains throughout the famine. These famines were typically followed by various infectious diseases such as bubonic plague and influenza, which attacked and killed a population already weakened by starvation.[20]

Historical Perspective

Historians have recognized that one of the major pre-requisites of the evolution of a civilization in any society is the surplus production of food grains and other agricultural produce.[21] All the world's early civilizations arose in low land river valleys where temperate climate and fertile, alluvial soils increased yields above the mere subsistence level. Only when food production has surpassed levels of bare subsistence, when agriculture is no longer a people's full time occupation, then the society has the time and energy to devote for the development in other fields like education, science and technology and for promotion of different welfare programmes like health, hygiene and cultural activities especially in the rural areas which characterize the growth of civilization. It is a universal fact that the development of agriculture and food production directly depends on the water resources, without adequate water resources, food production becomes scare and hence life is insecure. In the absence of perennial rivers, farmers totally depend upon the rainfall or irrigation facilities.

Unfortunately during the last several centuries many parts of our country have been prone to severe drought and famine conditions due to several factors namely political instability, internal chaos, spreading of severe infectious diseases like plague, cholera, etc. drastic change in climatic conditions and rapid deforestation.

Origin and Meaning of Famine

The origin of the word 'famine' can be traced to the Latin word 'fames' meaning hunger. 'Hunger' is an emotional word associated with fundamental bodily sensations which impel living things to seek for food in a multitude of curious ways. In fact famine is the aftermath stage of hunger, i.e. generally famine is something deeper than hunger.

Though the word 'famine' is a distressful concept but it is experienced in many ways. We have a plethora of definitions explaining the concept of famine. Encyclopedia of Social Sciences defines famine as "a state of extreme hunger suffered by the population of a region as a result of the failure of the accustomed food supply".[22] But this definition, which strongly emphasizes the failure of food supply in a particular region as a frequent cause to famine, is of no relevance in these days because of the tremendous development of transport and trade, which opened a new vent for the inflow or outflow of food items over riding the regional obstacles.

According to Encyclopedia of Britannica famine is an "extreme and general shortage of food causing distress and death from starvation among the population of a district or country".[23] In this definition the word starvation, which appears to be synonymous with hunger needs explanation. 'Starvation' can simply be defined as a suffering caused by lack of food. It occurs in a great variety of circumstances and has numerous causes. Sometimes starvation may be self imposed one as in the case of "hunger strike" or political agitation, etc. This type of starvation, which will artificially prevail, i.e. man-made, cannot be taken as for famine, because famine suggests the action of something supra power than that of man in its origin.

It has been envisaged above that famine in a region is due to scarcity of food caused by natural or man-made phenomenon resulting in human sufferings.

Natural factors are droughts, floods, soil erosion, crop-pests and diseases destruction of crop by birds, animals and unconditioned and irregular monsoons and cyclones. Man-made phenomena are lack of purchasing power due to unemployment, lack of alternative occupations lack of necessary skills, prevalence of unwanted and outdated skills. Low productivity in agriculture as well as in other areas due to underdevelopment of technology were due to lack of infrastructure facilities. Decline of cottage and small industries institutional and social barriers concentration of economic power and lack of knowledge about the available resources, etc.

Famines are temporary, but their effect is secular. If one is interested to state the salient features of the concept of famine, it can be said that:

1. Famine is a regional concept, its scope is limited,
2. The effects of famine are dreadful,
3. Famine is due to shortage of food articles either created naturally or want only,
4. Famine mainly springs up from agricultural sector,
5. Famine does not affect the highly affluent societies. It is a distressing problem for the under developed countries,
6. Famine is the result of underdevelopment of infrastructure facilities. With the development of trade and transport, the scarcity has been diminished.

Famines in Nineteenth and Twentieth Centuries

Numerous famines occurred in India throughout the nineteenth century, and their victims were often counted in millions. There is a fair amount of agreement among nineteenth century analysts and later economic historians concerning the proximate causes underlying these catastrophes. In most (nearly all) cases, famine followed massive crop failures resulting from drought. The immediate effect of these crop failures was not only to reduce food availability in the affected region, but also (more importantly) to shatter the rural economy. In particular, landless agricultural labourers found little employment as field activity was brought to a standstill while general impoverishment simultaneously enlarged the supply of casual labour, food prices increased. As the less vulnerable groups strived to maintain a reasonable food consumption level (possibly by selling assets), while trade was often slow to move food to the affected area from other regions. Wages lagged behind price increases, further aggravating the plight of agricultural labourers. The operation of the so-called "moral economy" did little to mitigate their sufferings, which all too often ended only in death. Thus severe famines frequently took place even when crop failures were only localised (as well as short-lived) and food was far from wanting in the country as a whole.

This recurring scenario was aptly summarized by Baird Smith's well known statement to the effects that famines in India were "Rather famines of work than of food", The same verdict was arrived at later by successive Famine Enquiry Commissions, as well as by most independent analysts-though there has predictably been much more controversy about the persistent causes of poverty in the same period. The Famine Commission Report of 1880 (the first major report of its kind) is with quoting her, not least because it unwittingly provides a fine blend of class analysis and modern "entitlement theory".

Another important aspect of the received analysis of 19th century famines is the view that entitlement failures occurred amidst plenty rather than in the context of a fierce battle for scarce food, There was a rather striking degree of agreement on this question in the official reports of the time, as well as among later commentators of very diverse persuasions from Baird Smith (1861) to Ghose (1982) and including Naoroji (1900), P.C. Ray (1901), S.C. Ray (1909), Loveday (1914), Srivastava (1968), Mc Alpin (1983) and Guz (1987) among many others. The literature on nineteenth century famines is replete with statements such as the following.

"What does a drought mean?" It is not a question of food the scarcity of food in a district affected by drought is the least of the evils with which the Government of India have to deal. There is nearly always a sufficiency of food in India to feed all the people within its limits and owing to the development of the railway, the British Government was able, no matter what part of the country may be affected, to pour in sufficient food to maintain the people of the district.

Painstaking investigations would be required to ascertain whether or not the absence of a food availability problem during nineteenth century famines in India was a myth propagated by the Famine Commissions. There is, however, little indication that the myth theory should be taken very seriously. The fabrication of a myth would have triggered dissent, of which there is very little trace. Indeed, it is rather striking that commentators of all persuasions, including many radical ones, concurred with the views of the Famine Commissions on the issue of food availability, the distinction of the radical writers was not to challenge the notion of sufficiency of food, but rather to trace the lack of purchasing power of the masses to colonial exploitation.

As far as quantitative evidence is concerned, several independent rounds of food availability calculations were carried out (in 1878, 1898 and 1902), they were based on food consumption allowances which ware extremely generous by contemporary standards and yet they all arrived at the firm conclusion that a substantial surplus 12 of food was available in India in normal years. Since multi-year storage was a widespread practice in nineteenth century India, and most famines were localised, the argument against the existence of a problem of physical availability of food seems convincing enough.

Indicators to Identity Famine Spot

The following indicators identify famine areas:

1. Failure of crops due to unforeseen weather conditions or any natural phenomenon.
2. Incidence and pattern of rainfall.
3. Percentage of irrigated land to total area and the availability of irrigation facilities.
4. The frequency of famine or scarcity in that area.
5. Revenue remissions.

The first indicator of the impending disaster is apprehended with the failure of crops. Crop failures are due to any natural factors like drought, flood, etc. There are two main types of crops in India Kharif and Rabi. Kharif requires more rains than the Rabi. The kharif crops are such as rice, millets maize and Rabi are like wheat, barley and oats. Rice is a monsoon crop and failure or shortage of the summer monsoon. Therefore, at once reduces the yield of the rice crops and the food supply over a large part of the country. The crop failures are more in dry regions where the average annual rainfall is less than 37.5 cm. Timely rainfalls is more important for the success of crop production in any region. Unseasonal rainfall also proves destructive to crops and produces famine conditions even in high rainfall areas. The tracts where the annual rainfall is less than 75 cm become liable to famines.

The second indicator is incidence and pattern of rainfall in the region. Rainfall in India is dependent in different degrees on the southwest and northeast monsoons on cyclonic depressions and disturbances and on violent local storms which occur in regions where cool humid wind from the sea meet hot dry winds from the land and occasionally reach cyclonic dimensions. As most of the rainfall in the country is received during southwest monsoon period. The pattern of annual rainfall resembles the pattern of monsoon period. It generally exceeds 100 cm in areas to the east of longitude 7^{0}E. It extends to 250 cm along almost the entire West coast and Western Ghats and over most of Assam and sub-Himalayan West Bengal.[24] The west of the Isohyetal joining Porbandar to Delhi and then to Ferozepur, rainfall diminishes rapidly from 50 cm to less than 15 cm in the extreme west. The peninsula has on elongated area of less than 60 cm with pockets of 50 cm.

The Southwest Monsoons

Due to the rapid rise in temperature in May over the Asian mainland, there is a corresponding drop in air-pressure over the area. June and July are crucial months for agriculture, and the fate of the kharif crop depends very largely on the distribution, and on the amount of rain, which the southwest monsoon brings to the land in these two months. The monsoon in its season does not continuously rain in any part of the country. There are alterations of heavy to moderate rain, and partial or general breaks.

The Northeast Monsoon

By the middle of October the belt of low pressure is transferred to the centre of the Bay of Bengal. The retaining monsoon winds cause occasional showers on the east coast, the amount of rainfall decreasing from the coast towards the interior. The districts of the Deccan and Tamil Nadu receive rainfall in these months almost entirely because of these cyclonic storms or depressions and hence the distribution of this rainfall is irregular, both in incidence and in duration.

Two indicators are used to know the incidence of drought, viz. (1) minimum rainfall of 750 mm per annum and (2) any area where 30 percent or more of the cultivated area is irrigated. On the basis of these two indicators the Irrigation Commission (1972) identified the following districts and taluks as affected by drought. Calculated the incidence of drought in terms of area and population. The table indicates the incidence of drought in India.

Table 1 Area and Population affected by Drought

S. No.	*State*	*No.*	*No.*	*Geographical area*	*Population (1000 persons)*
1.	Andhra Pradesh	7	60	9,700	9,410
2.	Gujarat	11	60	7,070	5,480
3.	Haryana	3	6	810	1, 490
4.	Madhya Pradesh	9	24	4,090	3,070
5.	Maharashtra	9	45	6,250	6,930
6.	Mysore	12	88	10.350	11.580
7.	Rajasthan	9	19	7,480	2.760
8.	Tamil Nadu	7	24	3.980	7.360
9.	Total	67	326	49,730	48,080

1981 Census, Source: *The Irrigation Commission Report* (1972), Government of India, New Delhi, Vol. 1. p. 166.

Table 1 reveals that the incidence of drought in Mysore is higher, in terms of area and population. The second place goes to Andhra Pradesh and followed by Tamil Nadu, Maharashtra and Gujarat. The incidences in southern states are very high when compared to northern states of India. According to Irrigation Commission (1972), the hard-core areas of drought comprise about 16 percent of the total geographical area of the country and account for over 11 percent of its population.[25]

The available irrigational facilities in the region are also one of the indicators of the famine spot. Artificial irrigational facilities are essential for the success of crops in the areas of deficient rainfall or when rains fail to provide sufficient water for growth. Irrigation works like projects tanks, canals and wells of the region is more important to give protection against failure of the rains. Which brings water to the fields of the farmer? According to Irrigation Commission (1972) while it is difficult to suggest any absolute percentage of irrigation which could be considered sufficient for drought areas, judging from the criteria followed in the past we feel that it would be safe to consider any taluk or equivalent unit, where 30 percent or more of the cultivated area in irrigated, as having reached a stage which would enable it to sustain a reasonably stable agriculture and to be reasonably protected against drought.[26]

The frequency of famines or scarcities is identified by the Irrigation Commission on the basis of the probability of critical rainfall shortage of 20 percent. Using the annual and southwest monsoon rainfall data from 1901 to 1960 for about 500 stations, which are fairly representative of the drought and chronically drought areas.

If crops were damaged or completely destroyed in any year due to natural calamities either the amount of tax is reduced or a total remission is granted. Remissions and suspensions of payment of land tax, when crops fail had long in addition to the absence or the severe deficiency of rainfall over a fairly long period. Several factors such as precipitation, temperature, wind velocity, sunshine, soil texture, soil moisture and antecedent rainfall interact to out give drought situation. However, the key role is played by rainfall and the crucial variables are its distribution, its variability and its capacity to meet evapo-transpiration needs. This is a wide and comprehensive definition of "drought". This has listed out several casual factors leading to drought situation the supreme place among them being given to rainfall.

In 1874 the response from the British authorities was better and famine was completely averted. Then in 1876 a huge famine broke out in Madras. Lord Lytton's administration believed that 'market forces alone would suffice to feed the starving Indians.' The results of such thinking proved fatal (some 5.5 million starved),[27] so this policy was abandoned. Lord Lytton established the Famine Insurance Grant, a system in which, in times of financial surplus, INR 1,500,000 would be applied to famine relief works. The result was that the British prematurely assumed that the problem of famine had been solved forever. Future British viceroys became complacent, and this proved disastrous in 1896,[28] about 4.5 million people were on famine relief at the peak of the famine.

No doubt we had no 'famine' like those in the previous centuries even in the worst drought year but still large numbers of people suffered from hunger, malnutrition, misery and deprivation. Some States like Andhra Pradesh, Rajasthan, and Gujarat. Orissa and Karnataka had the misfortune of having successive droughts in the present century too, even after independence. Individual desperate, cruel and deplorable cases of sale of children for a few rupees and some clothes were reported from a few districts of Andhra Pradesh and Orissa, though denied by the state governments. The Prime Minister, Sri Rajiv Gandhi, had visited the district Kalanidhi of Orissa and some make shift arrangements were made to give a face-lift to this district and other areas visited by the Prime Minister in this state and elsewhere. But the fact of the matter is that children, old persons and infirm did die in hundreds if not in thousands, due to malnutrition and eating of harmful forest products. As for cattle, thousands died in Rajasthan alone and thousands migrated to greener pastures or even to cities including Delhi.

A grim though factual picture based on field visits has been given in the magazine India Today of September 15, 1987 under the heading 'The Devastating Drought'. This magazine has again highlighted this problem in its issue of 15th May 1988 captioned "Gloom despite Boom". A record rabi harvest but reserves are still low. An extract is given below:

> Their latest estimates put the production at an all time record of 66 million tones, wheat, the main rabi crop was likely to touch a peak 47 million tones but gloom was due to the fact that the stocks of food grains with the F.C.I. dropped from 23 million tones to worrying 11 million tons last week. This

> has compelled the government to import one million tons of wheat and one official in the Food Ministry has said. We will import more as and when necessary, the latest news is that the government is going to import half a million tons of rice from Thailand.

The question arises as to why do we have this crisis? Have our objectives of the Five Year Plans, of self-sufficiency in food grains, by removal of poverty, removal of unemployment and of provision of minimum need failed completely or partially? It is obvious that the Green Revolution has not reached many parts of India particularly in the eastern region where productivity of rice. Its main crops are still very low. The other three objectives have met with a partial success only so far as would be clear from drinking water crisis in summer. Even in cities like Delhi, Mumbai, Bangalore, Hyderabad, Chennai, etc. what to say of rural areas where millions of women and children have to go kilometers and spend 6 to 8 hours daily in fetching drinking water, fuel and fodder for their use and cattle. Still their cattle had died by thousands and they particularly children, old and infirm have suffered untold misery.

Despite huge buffer stock of food grains, millions, particularly weaker sections, like small and marginal farmers landless agricultural labourers, village artisans, tribes and urban poor do not have the purchasing power to purchase food and other necessities of life, enough for their families to have at least one square meal a day for want of adequate employment opportunities within easy reach of their homes. This situation resulted in massive migration to towns in search of work, particularly in severe drought years adding to creation of more slums in big towns and breakdown of civic amenities quite inadequate even otherwise.

Sustainable Development

However, India has made significant progress in the post-independence period through Green Revolution by tripling the food grain production, doubling the cotton and sugar production. However, adoption of intensive agricultural practices has led to problems like water logging soil salinity, nutrient deficiency soil erosion, etc. in large tracts. Added to this, recurrent natural disasters like droughts and floods impose serious constraints to natural resources development. Further, the population explosion coupled with urbanization and industrialization and the resultant decrease in arable land available for agriculture have put tremendous pressures on the land and water resources as well as environment, while there is no option but to produce more food and achieve necessary socio-economic development. It is important to recognize that today's economic progress should not be at the expense of tomorrow's developmental prospects. Development must be sustainable over time. Sustainability has many dimensions ecological, economic, social, and cultural that has to be suitably integrated, with the environment to develop a cost-effective, energy-efficient, environmentally benign management system. It needs application of the best in modern technologies such as space technology, biotechnology and information technology.

Sustainable development of natural resources relies on maintaining the fragile balance between productivity functions and conservation practices through precise identification and systematic monitoring of problem areas in various resources and developmental sectors. Further, it calls for application of alternate agricultural practices, crop rotation, use of bio fertilizers, energy efficient farming methods and reclamation of under-utilized land, wastelands, planned exploitation of mineral and ground water resources, etc.

All these leads me inevitably to a brief evaluation of one of the severely' drought prone area of Andhra Pradesh, i.e. Anantapur district the area of my study concerning drought, its gaps and weaknesses.

The Concept of Drought

Drought is one of the most serious natural hazards to mankind. They have been the root causes of famines in many parts of the world and more so in our country where agricultural success and prosperity are virtually dependant on the amount and distribution of rainfall. Drought is "a meteorological phenomenon and the vulnerability of any area to drought depends on the extent to which physical and climatic conditions play an adverse role in creating an unstable agriculture. Drought is actually a period of scarcity of water, which results in less production of crops. In economic sense, it is a period of intense economic disaster resulting from unemployment, loss of agricultural production and acute shortage of water and pastures for animals and a decline in food cropped area. B. Arunachalam has rightly pointed out water is at the heart of Indian economy sustaining its teeming millions and often paralyzing the slowly turning wheels of the economy.[29] In India agriculture is the primary sector, which depends more on rainfall for water resource supply. The success and failure of agriculture is heavily affected by the performance of monsoon. Therefore, if any drought occurred in India history spells out that it is to a heavy extent due to shortage of rainfall.

According to the Irrigation Commission (1972) "Drought is the result of an imbalance between the soil moisture and evapo-transpiration needs of an area, over a fairly long period. So as to cause damage to standing crops and a reduction in crop yield. The basic characteristic of drought is a steady rise in temperature".

According to Encyclopedia of Britannica "Drought results from long-continuous dry weather and lack or insufficiency of rain which causes exhaustion of soil moisture". Suffering of plants from lack of water depletion of underground water supplies and reduction and eventual cessation of stream flow.[30] The meteorological department has defined drought as a situation occurring in any area when the annual rainfall is less than 75 percent of the normal.[31] It has defined 'moderate drought' as obtaining where the rainfall deficit is between 25 to 50 percent. This definition attempts to quantify the degree and extent of drought. The problems posed by drought vary from area to area depending on the amount of rainfall and its variability, and also the extent to which irrigation has been developed. Assuming that districts, which receive less than 750 mm of rainfall per annum are liable to drought. The British Meteorological glossary speaks that "a climate in which the rainfall is insufficient to support vegetation is termed as arid and drought is dryness due to lack of rain".[32] The most important key point of all these definitions is deficiency of rainfall. There is no widely accepted definition of drought. When the shortage of moisture will affect the established economy of a region, then only it can be called as drought. If this situation becomes a frequent occurrence in any area, e.g. two years of poor rainfall followed by a third year without any rain at all, it is then that famine makes its appearance in the region. Thus, the most important natural cause of famine has been drought. Most of the great famines of history have been precipitated by drought India receives 3,700,400 cubic inches of rain annually on an average. But there is a variation in the seasonal and annual distribution of rainfall. Nearly 12 percent of the country's area receives less than 71 cm of rainfall and only 8 percent of the country receives more than 254 cm. Some parts of Rajasthan get as low as 12.70 cm of rain, while hills of Assam 1.1 cm as much as 1.270 cm. This sort of deviation in rainfall is the major cause of drought in India.

Not only low rainfall, but also untimely distribution of high rainfall leads to droughts. Most of the rainfall, say 80 percent in generally received during the period from June to September in most parts of the country.

There are 128 districts in India with 68 million hectares of land having low to medium rainfall, i.e. below 1.125 mm annually and very limited irrigation facilities. Of these 128 districts, 25 districts are characterized by very poor rainfall in the range of 375-750 mm per annum with an irrigated area of 5-10 percent of the cropped area. Another 91 districts receive annual rainfall from 750 mm to 1.125 mm. The total net sown area in these districts is 42 million hectares of which about 5 million hectares are irrigated.

Factors Determining the Incidence of Drought

It is an important issue to determine the incidence of drought, i.e. to determine whether any area is drought-prone or not various authors to determine the incidence of drought have adopted several methods. In this study three of those methods have been considered.

- Statistical techniques
- Non-statistical methods
- Water-budget methods

Statistical techniques are mainly based on shortage of rainfall either on a weekly, monthly or yearly basis. Non-statistical techniques are based on climatic data such as temperature and atmospheric humidity. Statistical and non-statistical methods indicate only the atmospheric drought being more tabulations of days receiving less than a specified amount of rain and the other climatic data do not take into account the important part played by soil moisture on which the growth and development of crops depend. They do not give any information about soil drought as it affects crops. Soil drought begins when available moisture in the soil is diminished to the extent when crops no longer obtain moisture rapidly enough to replace transpiration loss. Thus, statistical and non-statistical methods suffer from severe limitations.

Water budget method which considers rainfall, soil moisture storage and water needs of the crop day after day during the entire crop period, gives an exact picture of agricultural drought. The frequency of drought days will depend on the rainfall pattern, the moisture storage capacity of the soil, the rooting depth and evapo-transpiration rate. With the help of water-budget method, using long time meteorological data soil and crop data, drought-affected areas can be identified.

Occurrence of Drought and its Probability

Drought, which occurred as an interaction between agricultural activity and natural events, resulted in a water vacuum or inadequate supply of water for plant and animal needs. It is too difficult to recognize the beginning of drought, but the end of drought is not such a difficult task to recognize. If rains occur, it indicates the end of drought. There is no particular method to forecast the occurrence of drought and its probability.

Distinction between Famine and Drought

After having pursued the concepts of famine and drought, it is worthwhile to know the distinction between these two concepts, the main distinction is one of casual factors. Famine may be caused by any natural or man-made phenomenon. The natural causes are like drought, which occurs due to shortage of rainfall or uncertainty or unseasonal rainfall, floods, crop diseases, soil erosion, etc. man-made or artificial causes are like lack of purchasing power, unemployment, lack of industrial development, lack of infrastructure, excessive land assessment decline of village and cottage industries and defective land utilization, etc. Thus both may cause by natural or artificial factors or famine. But drought is a meteorological phenomenon which depends upon rainfall. Drought occurs due to low rainfall or untimely rainfall.[33]

One important aspect of these two concepts is that one is imbibed in the process of the other. One is the precedent to the other. Drought precedes famine and famine is inherent in the process of drought. Most of the great famines in history have been caused by drought of course, there may be equal consequences of famine and drought, but there is difference between these two concepts.

References

1. Nash, J. Madeleine (2003), *El Niño: Unlocking the Secrets of the Master Weather-Maker*, London: Warner Books.
2. Collier, Michael and Webb, Robert H. (2002), *Floods, Droughts, and Climate Change*, Tucson: University of Arizona Press.
3. Dreze, Jean (1991), 'Famine Prevention in India', In Drèze, Jean and Sen, Amartya (eds.), *The Political Economy of Hunger: Famine Prevention*, Oxford: Oxford University Press, pp. 32-33.
4. Currey, B. and Hugo, G. (1984), 'Famine as a Geographical Phenomenon', *Geo Journal Library*, Boston: D. Reidel.
5. Singh, N. (2002), *Population and Poverty*, p. 112.
6. Visaria, Leela and Visaria, Pravin (1983), 'Chapter V: Population (1757-1947)', In Kumar, Dharma and Desai, Meghnad, *The Cambridge Economic History of India*, Cambridge: Cambridge University Press, pp. 463-532.
7. James, Lawrence (2000), *Raj: The Making and Unmaking of British India*, Macmillan, pp. 49-52.
8. Bowen, H.V. (2002), *Revenue and Reform: The Indian Problem in British Politics 1757-1773*, Cambridge: Cambridge University Press.
9. Desai, Meghnad; Raychaudhuri, Tapan and Kumar, Dharma (1983), *The Cambridge Economic History of India*, Volume 2, Cambridge: Cambridge University Press, p. 477.

10. Currey, B. and Hugo, G. (1984) *op. cit.*, p. 75.
11. Gourlay, J. (2003), *Florence Nightingale and the Health of the Raj, Aldershot, Hants*, England; Burlington, VT: Ashgate.
12. Srivastava, H.C. (1968), *The History of Indian Famines from 1858-1918*, Agra: Sri Ram Mehra and Co. Sen, Amartya (1982), *Poverty and Famines : An Essay on Entitlements and Deprivation*, Oxford, Clarendon Press. Bhatia, B.M. (1985), *Famines in India: A Study in Some Aspects of the Economic History of India with Special Reference to Food Problem*, Delhi: Konark Publishers.
13. Mander, Harsh (2009), 'Barefoot: Colonial Legacy of Famine Codes', Chennai: *The Hindu*, retrieved 3rd October 2010, p. 1, Davis, Mike (2001), *Late Victorian Holocausts: El Nino Famine and the Making of the Third World*, New York: Verso Books, p. 299.
14. Davis, Mike (2001), *op. cit.*, pp. 299-300, Wong, J.W. (1998), *Deadly dreams: Opium, imperialism, and the Arrow War (1856–1860) in China*, Cambridge: Cambridge University Press.
15. Martin Ravallion (1987), 'Trade and Stabilisation: Another Look at British India's Controversial Foodgrain Exports', *Explorations in Economic History*, vol. 24, Issue 4, October, pp. 354-370.
16. Roy, Tirthankar (1999), *Traditional Industry in the Economy of Colonial India*, Cambridge University Press, p. 18. Roy, Tirthankar (2006), *The Economic History of India, 1857-1947*, p. 362.
17. Currey, B. and Hugo, G (1984), *op. cit.*, p. 91.
18. *Ibid.*
19. Ghose, Ajit Kumar (July 1982), 'Food Supply and Starvation: A Study of Famines with Reference to the Indian Sub-Continent', *Oxford Economic Papers*, Oxford: Oxford University Press, vol. 34, no. 2, p. 380.
20. Walsh, Judith E. (2006), *A Brief History of India*, pp. 144-145.

21. Aykroyd, W.R. (1974), *The Conquest of Famine*, Chatto & Windus, London, p. 2.

22. South Ward, 'Famine', *Encyclopedia of Social Sciences*, Vol. VI, p. 85.

23. *Encyclopedia of Britannica (A Survey of Universal Knowledge)*, vol. 9, p. 63.

24. Sir James, A. Murray (ed.), *Palgrave's Dictionary of Political Economy*, p. 19.

25. Thomson, W.S., vol. 54, p. 735.

26. *Report of the Irrigation Commission* (1972), p. 28.

27. *Ibid.*, p. 166.

28. Gilmour, David (2007), *The Ruling Caste: Imperial Lives in the Victorian Raj*, New York: Macmillan.

29. *Ibid.*, p.165.

30. Arunachalam, B. (1971), *The Problems of Drought. What crop the Geographer do'r Proceedings*, Symposium on Revenue Development Planning, Madras, 5th-7th January, 1971.

31. *Encyclopedia of Britannica*, vol. 7, p. 699.

32. *The Irrigation Commission Report*, 1972, p. 160.

33. *Meteorological Glossary*, Meteorological Off., London, 1957. p. 17.

Chapter III

Famines and Economic Deprivation of Agrarians

Famine has been defined as widespread food shortage leading to a significant rise of the regional death rate.[1] The demographic component of this definition is clear famine is a disaster characterized by large numbers of excess death so The strength of this definition however is also its weakness for simplicity often obscures more than it reveals. Famine in fact is a complex syndrome of multiple interacting causes, diverse manifestations, and involving all three demographic variables - mortality, fertility, and migration. A sound understanding of the complex mechanism by which disaster overwhelms the social, economic, and demographic stability of a society is essential not only for effective remedial action but also for long-term prevention.

Famine has been defined as "a state of extreme hunger suffered by the population of a region as a result of the failure of the accustomed food supply."[2] This description of the calamity is valid only under primitive and mediaeval conditions of economic life. Due to lack of means of transport and absence of well-established channels of trade, man under those conditions had to subsist on food raised either by himself or by others in his immediate neighborhood. The people of a region, in the event of a local failure of crops, cannot look to other parts of the country, much less beyond their national frontiers, to make up the deficiency in their current food supply.

Famine under these conditions, therefore, means absolute want of food in the affected region and the high and the low all suffer alike from its ravages when it occurs.

Modern developments in industry, trade and transport have wrought a radical change both in the meaning and nature of the famine problem. Instead of absolute want, famine, under modern conditions, has come to signify an abrupt and sharp rise in food prices which renders food beyond the reach of the poor who suffer starvation. In a modern famine, food may be available at all times in the market but prices are so high that the poor people cannot purchase it.

Before the industrial and commercial revolutions occurred in Europe, famine was a natural calamity from which no part of the world was completely immune.[3] There was no direct relation between the economic conditions of a people and their vulnerability to famine. Thus, some of the most terrible famines known in World history occurred in ancient India, Egypt,[4] China, Greece and Rome at a time when these were economically the most advanced countries of the world. On the other hand, in Inter times, Britain and countries of Western Europe, the economically backward and poor, were more frequently menaced by famines than any other part of the globe.[5] But then, from the seventeenth century, Europe began to forge ahead, at first in commerce and then in industry. By the middle of the nineteenth century, it had already become the industrial and commercial leader and in material prosperity was the most advanced part of the world. Famine was almost banished from Europe after 1850, but mean while Asia passed under its sway. Under the new conditions, famine ceased to be a natural calamity and was transformed into a social problem of poverty and dearth. It did not affect the rich nations or even the richer sections of the poor nations, but became peculiarly a problem of the underdeveloped countries, where, even today, the larger portion of the population lives in a state of chronic hunger and under-nourishment.[6] A direct relation was thus established between poverty and the incidence of famine.

This transformation in the nature of famine was wrought in India by the construction of railway lines and development of trade after 1860. At the same time, there were other forces at work leading to steady impoverishment of certain classes, thereby creating a class of potential victims of dearth and scarcities. The ground was thus prepared not only for the change in the nature of famine but also for increase in its frequency. The study of famines in India after 1860, therefore, involves an analysis of the changes that occurred in the Indian economy and the consequences that they produced for different sections of the country's populations.

Food Crops in India

The people of India subsist for the most part on vegetarian food, consisting of cereal grains, pulses, and vegetables, supplemented sometimes with dairy products. Consumption of animal food such as meat and fish is restricted to small sections of the population. The grain crops, which are the principal means of subsistence, are of crucial importance. However, there are wide variations in the types of food grains raised and consumed in different regions. Rice is the staple food of the people of those parts of the country where climatic conditions are suitable for rice culture, such as Bengal, Assam, Bihar and Orissa, the costal and southern districts of Madras, and the western districts of Bombay. Considerable areas in the uplands of India are dependent upon millets, mainly jowar and bajra. All these crops known in India as kharif harvest are raised in the summer months. They constitute the principal source of food supply of large sections of the population. There is, however, another crop known as Rabi which is more characteristic of the parts of the country which have the coldest winter, the period of maturity of which covers winter months and which is harvested in the spring or early summer months. The principal grain obtained from Rabi is wheat. Famine in India, in the past, has usually resulted from the failure of the kharif harvest though in those areas where Rabi is important, the failure of the latter, following upon that of the former, adds immeasurably to the severity of the distress.

Rainfall and Crop Production

In a country of tropical climate with its cultivated area for the most part unprotected by artificial irrigation, rainfall must play a crucial role in determining crop yields and supplies of food from land, that indeed is the case in India. The immediate cause of famine in this country is almost invariably drought or unseasonal rains.[7]

The principal source of rain in the Indian sub-continent are the two periodic winds, known as the summer and the winter monsoons. The former gives rainfall in the eastern, northern and southwestern parts of India from June to October, and in Caryatid and Madras from October to December the latter brings limited, but agriculturally very beneficial, rainfall to the Punjab, Uttar Pradesh, Central India and Rajputana in the winter months of December and January. About 90 percent of the total rainfall in the country is obtained from the summer monsoon.

The regional distribution of rainfall in the sub-continent is uneven and shows wide variations. Thus at one end of the scale we have two zones of excessive rainfall comprising the Western Ghats and the outer slopes of the Himalayas with the hills of East Bengal and Cachar, nearly the shores of Assam, the plains of Cachar, Sylhet and the eastern fringe of the Gangetic Delta. The annual rainfall in these regions varies from 70 to 200 inches,[8] parts of eastern India not included in the zone of excessive rainfall, Orissa, and the Gangetic plain as far west as Kanpur, and beyond this, a belt varying from 50 or 60 to 100 miles in width, which skirts the base of the Himalayas up to the further extremity of the Punjab, are included in this region. It also includes the whole of Central Provinces,[9] the Plateau of Bundelkhand and Malwa, the eastern half of Hyderabad, the Eastern Ghats and the Coast plains of Carnatic with a narrow belt on the summit of the Western Ghats. The rest of the peninsula has a rainfall of less than 30 inches and in certain limited tracts as low as between 10 and 15 inches only. This dry region includes the eastern half of the Deccan plateau and Hyderabad, Mysore, and the plains of Coimbatore and Madura. Another equally dry region to the northwest includes Kathiawar, a large part of the western Rajputana, North Western Provinces (known later as United Provinces of Agra and Oudh and now Uttar Pradesh) around Agra and Delhi and a belt of less than 100 miles wide in the North Eastern Punjab. Finally, there is the remainder of Rajputana, Sind, and the southern half of the Punjab which has an average annual rainfall of less than 15 inches, indeed for the most part less than 10 inches.

The tract more especially liable to severe famines is that where the annual rainfall averages between 15 and 60 inches. Of this, the areas with the annual rainfall of less than 30 inches or so, suffer the most because of the greater frequency of drought in them.[10] Failure of rains is less common in the parts which have an annual average of 30 to 60 inches of rain. But when it occurs, it proves very destructive because, generally speaking the population in these regions is dense, the holdings are small, and the lower classes of areas, therefore, in the past, have been the seats of the most disastrous famines. Bihar, Orissa, West Bengal, the North Western Provinces, Oudh, Rajputana, Berar, the Northern Circars (Madras), Khandesh (Bombay), Hyderabad, Mysore and the Carnatic are all too familiar names in the history of famines in India in the nineteenth century.

A part from an absolute deficiency, unseasonableness of rainfall also proves destructive to crops and produces famine conditions even in the areas of fairly high rainfall.[11] In India, the choice and combination of crops, in each of the meteorological regions of the country, is nicely adjusted to the quantity and seasonal distribution of agricultural water supply. Vagaries of seasons, therefore, have far-reaching effects upon the fortunes of the cultivator. Thus, in the case of both dead food and aghani harvests of rice in the Lower and the Middle Ganges Valley, premonitory showers at the beginning of summer are important for ploughing the land and sowing the seeds. In the case of the former, rain is necessary again in September, while the latter needs, for purposes of transplantation, showers in October Failure of October rains (locally known as Hathia) spells disaster especially to districts like Bankura, Midnapore and Khulna, where more than 70 percent of the cultivated land is occupied by the winter rice crop which supplies the major portion of the annual food supply of their population. The precarious dependence of the people in these districts on a single crop for their food supply, which in turn depends upon timely rainfall, makes them the blackest spots on the famine map of Bengal.[12] In the Western and Upper parts of the Ganges Valley, the situation is slightly better in this respect for the farmer has, in these parts, a choice, depending upon the conditions of soil and climate, between the kharif and the rabi harvests, the former producing rice and millets and the latter wheat, barley and gram.

But as in the Lower and the Middle Ganges Valley, a nice adjustment between choice of crops and seasonal rainfall has been achieved in these parts as well. Thus in the eastern districts of what was formerly known as North Western Provinces, if rains fail in June and July, the sowing and sprouting of seeds for the kharif harvest is impeded and no amount of rainfall in September can revive the withered crop. The cultivator, disappointed by the failure, seats to work in September and October for the rabi crop. For the success of this crop, rainfall in December and January is of crucial importance. If the drought in June and July is followed by failure of winter rains also, the prospects for the cultivator become bleak.[13]

The seasonableness of rainfall is important for the success of crop production even in the areas of less copious rainfall, Thus, in the former Madras Presidency, Settlement Reports of the various ryotwari districts frequently speak of proper distribution of rainfall over various months being of greater importance to the cultivator than even the total amount of rainfall in the year, Mr. Crole, for instance, writing about Kurnool district, said "The amount of rainfall required for the whole year for a good crop in this district seems to be about 25 inches. If, however, the rains are regularly distributed, about 15 inches would be sufficient".[14] The popular saying is, and there can be no doubt of its correctness, Mr. Crole continues, "that one good fall of rain about an inch and a half in each fortnight after seed is sown, is sufficient for agricultural necessities". But the success of crop does not seem to depend so much on the quantity of rain which falls, as on its proper distribution over the different periods of agricultural operations and at different stages of the former Bombay Presidency by Dr. Harold Mann in a more recent study of rainfall and famine in these districts.[15]

Frequency and Nature of Famines

India has suffered from famines since time immemorial.[16] Though a connected and complete account of all the famines that occurred in the pre-British period of Indian history is lacking, the available evidence suggests that in the earlier times a major famine occurred once in every 50 years.[17] From the beginning of the eleventh century to the end of the seventeenth, there were fourteen famines almost all of which were confined to regions very much limited in area.[18]

The frequency of famine showed a disconcerting increase in the nineteenth century. In a period of about 90 years, from 1765 when the British East India Company took over the Diwani of Bengal to 1858, the country experienced twelve famines and four "severe scarcities."[19] The frequency of famine showed a still further increase during the first fifty years of the direct rule of India by Britain. Between 1860 and 1908, famine or scarcity prevailed in one part of the country or the other in twenty[20] out of the total of forty-nine years.

Even more interesting than the increase in frequency was the change in the nature of famine over the period under study. So long as the means of transport were scanty and trade in grain limited, famine meant extreme hunger faced by population of a region as a result of failure of the accustomed food supply caused generally by the climatic factor. With the development of the railways in the country, the famine problem underwent a radical change. In 1867, the Famine Commission which enquired into the famine of 1866 in Bengal, Bihar and Orissa was emphatic in defining the word "famine," in "its ordinary and popular acceptance" as "suffering from hunger on the part of large classes of the population."[21] During the scarcity of 1891 in Bengal, however, the Lt. Governor of the province described, "the present state of things" by saying that "there is apparently food in the country and that anyone can feed himself for an anna a day, but the usual agricultural labor by which landless classes earn wages is mostly at a standstill, and they have to resort to relief works to earn that anna."[22] The scarcity in Bengal in 1891 was a limited local affair. Even during the Great Famine of 1896-97 which was a countrywide calamity, food was "always purchasable in the market though at high and in some remoter places at excessively high prices".[23] This was in spite of the fact that shortfall in the production of food grains in the country that year was estimated at 18 to 19 million tons.[24]

The conclusion has often been drawn from this that the nature of famine in the latter half of the nineteenth century had changed from a shortage of food supply, as in the past, to lack of purchasing power with those who suffered starvation. Strictly speaking, this is not true. The development of means of communication and transport helped to distribute a local scarcity evenly over the country as a whole resulting in rise in prices of food grains which rendered food beyond the reach of poorer sections of the community. The abnormal rise in food prices during years of drought was itself a measure of the shortage of food grains in the country as a whole. Thus, instead of absolute lack of food in one region famine under the new conditions assumed the form of a sharp rise in food prices. The process was helped by the emergence of "destructive" from of speculation in food grains[25]. The disappearance of domestic stocks, which people were accustomed to keep in the past, and the development of a large export trade in grain, which siphoned off annually a substantial part of cereal production of the country, were factors which contributed to the rise in prices of food grains during the period of famine. As a result of these developments, the rise in prices and the consequent suffering of the people were out of all proportion to the natural scarcity.[26] The human and institutional factors were becoming more important than the natural scarcity in causing distress and starvation. Even as early as 1861 Baird Smith noted the feverish speculative activity in grain trade that was going on during the famine in North Western Provinces as also the strange phenomenon that "prices had risen, on account of active export trade, to the famine levels even in districts having most bountiful corps." Similarly, Colvin pointed out, during the same famine, that "where, there is canal irrigation in Pargana Muzaffarnagar the crops are thriving but the people are starving on high prices."[26] This was before the advent of railways. After lines had been laid, speculative holding of stocks by traders emerged as the most important contributory factor to the causing of distress arising from natural scarcity.

The then Famine Commissioner Sir William Hunter has stated[27] in his "Annals of Rural Bengal" that the whole tendency of the modern civilization was to rear up "intervening influences" which rendered the relation of actual pressure to natural scarcity less certain, and less direct, until the two terms, "which were once convertible, come to have very little connection with each other." This is what happened in India in the latter half of the nineteenth century. Under the modernizing influence of railways and the growing grain trade the famine problem was transformed from one of intense local sufferings limited to the area of severe drought to that of high prices of food over a wide area accompanied by the inability due to poverty of a large mass of the agricultural population to feed them.

Famine and Poverty

The mortality statistics of famines are silent about the class composition of the victims of starvation but information about the classes of people who came to seek government relief is available. Thus in 1860-61, while detailed statistics were not given by him, we have it on the authority of Baird smith who made enquiries into the circumstances of the famine of that year that "common inmates of out relief houses are the ordinary agricultural labourers." As regards castes, Baird Smith writes in another part of his Report, "By far the largest portion was from the common village labourers, next came the native weavers, while all other castes were represented by me fraction of the whole numbers."[28]

As regards sufferings of native weavers, Baird Smith point out that "among the sufferers none are more commonly met with or more helplessly prostrated that the great body of the native weavers," and "next to ordinary agricultural labourers, they are the most common inmates of our relief houses."[29] The Famine Enquiry Commission of 1867 maintained that the principal victims of famine were the agricultural wage labourers and that, under the new conditions created in the country by the British administration, "there is reason to expect a gradual increase in the classes who may hardly withstand a scarcity not amounting to that extreme famine which involves the whole population."[30]

The most interesting evidence on the point, however, is furnished by the census of 5,000 labors on relief works, taken by Mr. James White during the famine of 1869-70 in the Bijnor district in North-Western Provinces.[31] According to this census, of the 5,000 relief workers 2,030 were Chamars "nearly all of whom are believed to have been filed hands, ploughmen and day labourers." On the other hand, Jats numbered 33, Rawas 34, Lodhras 3 and Rajpoots 4.These agricultural castes were either proprietors or cultivating tenants. Among the artisan class, the most predominant were the Hindu and the Mussalman "Jullahas" (weavers), the two together numbering 816. The number of other artisan castes was insignificant, Rangrez (dyer) 1, Barhai (carpenter) 7 and Dhunia (carder) 31. The inference drawn by Mr. Henvey from these figures was that "actual holders of land managed to subsist without external aid" and that in future "Government may leave almost entirely out of calculations the great agricultural classes and confine its attention to the industrial and laboring classes. " In the event of famine "what should be chiefly looked to," Mr. Henvey continued, "is the state of the labour market."[32.]

Among the "great agricultural classes" the small cultivator managed to tide over the period of scarcity without resort to relief camps only with the help of credit from the moneylender. When credit failed or when his economic condition had already been gravely undermined and his credit exhausted in consequence of a previous calamity, he had no alternative but to rush to relief houses. This fact was brought out by the remarkable increase in the number of cultivators in relief camps in the Bombay Presidency during the famine of 1900. A census of classes and castes of people employed on relief works in the Presidency disclosed the following interesting facts.[33]

Table 1 Classes of People Seeking Relief

	GUJARAT					
	Ahmadabad		Kaira		Panch Mahal	
	Number	percent	Number	percent	Number	percent
Cultivators	48,889	66.4	11,086	67.7	7,133	41.9
Labourers	2, 083	2.8	442	2.7	618	3.7
Depressed Classes	4,611	6.3	3,007	18.4	69	0.4
Weavers	670	0.9	104	0.6	82	0.5

contd....

DECCAN								
	Khandesh		Ahmednagar		Poona		Sholapur	
	Number	percent	Number	percent	Number	percent	Number	percent
Cultivators	67,452	45.7	39,562	40.5	9,872	50.4	55,181	40.1
Labourers	2,188	1.5	4,483	4.6	548	2.8	2,053	1.5
Depressed Classes	41,841	28.3	33,568	34.4	6,870	35.1	47,520	34.6
Weavers	696	0.4	676	0.7	9		552	0.4

The 'depressed classes' in the rural areas were, occupationally, mainly agricultural labourers. Adding up these two "castes" in the foregoing table, we reach the conclusion that in the famine of 1900 in the Bombay presidency the class of cultivators formed the large majority of the labourers on relief works and that agricultural labourers came next.[34]

It is in the increased vulnerability, during the period under study of the three sections of the rural population, namely, agricultural labourers, weavers and the tenant cultivators, that the explanation must be found for the greater frequency of famines and the increase in sufferings caused by them at a time when, on account of the presence of the railways which helped in the distribution of supplies, the danger of famine in the country should have receded.[35] This, in turn, is explained by a set of factors like the industrial decline of the country, the disintegration of the old social organization and institutions, which in the past had afforded protection to the individual against occasional misfortunes, and dispossession of the hereditary agriculturist of his property by the urban capitalist. The weavers suffered directly as the result of decline in their craft in the nineteenth century; labourors lost their customary ties with the proprietary class and gradually drifted into the position of day labourers and tenants were affected adversely by the intrusion of the urban capitalist in the person of rent receiving landlord and the extortionate moneylender into the economy of the village which, at one time, had presented a closed economic circuit. Besides, the caste and joint family systems as also the corporate village life had in the past, afforded to the individual a system of social security against occasional misfortune with the disintegration of these social institutions the destitute and the pauper were thrown for sustenance on public charity in times of distress.

In the final analysis, all this was the result of the fundamental changes that Indian society was undergoing and the powerful political, economic, and social forces that were operating at the time. The frequently recurring famines came in a period of rapid economic and social changes and the intensity of the suffering caused by them was immeasurably increased by the cumulative impact of those changes.

Causes of Poverty of Agriculturists

While superficially famine appeared to be merely a period of temporary dislocation of employment for the large mass of agricultural population, and abnormal rise in prices and failure of wages to keep pace with them, its causes lay deep in the economic and social changes that were taking place in the country. For an explanation of the increased vulnerability of certain sections of population to famine, we must therefore turn to the genesis of these changes.

The branch succeeded to a situation in India which[36] was very different from the one that obtained in their own county. England at that time was at the threshold of the Industrial Revolution which rapidly transformed her into a capitalist county with world leadership in trade and industry. The Indian society, on the other hand, with its base in the self-sufficient village communities, was still mediaeval in structure and organization. Merchant capitalism was, of course, well entrenched in the county and capitalist forms of organization had made their appearance in urban industries such as shipping and textiles. But the rural society appears to have remained largely unaffected by these influences. Till about the beginning of the nineteenth century the characteristic form of rural organization was the village community with its corporate life in which rights and obligations as also the economic and social relations between different sections of the community were regulated by custom and administered through a panchayat (village council), or by a headman who represented the collective will of the people in the village.

The village community dealt collectively with the State in the matter of payment of land revenue. Land in the village was transferable from one member to another but it could not be sold or transferred to anyone outside without the permission of the village community.[37] Only in rare cases was an outsider permitted to acquire land in the village and get admission to the membership of the propriety body of the village. Again, each village community had its dealing with a banker and a shopkeeper or merchant but the latter, if not subservient, was at best the ally of the Proprietary body[38] and he could not be regarded at that stage of the county's economic development an exploiting agent in the rural economy. The artisans and the class of menial servants might be regarded as the employees of the proprietary body of cultivators but the basis of their engagement with the cultivating community was the assignment to them of a fixed share in the produce of the land in exchange for their services, and not a contractual piece payment in proportion to the work done by them. The whole organization was thus adapted to the requirements of a corporate form of living centered on the cultivation of land for the production of food grains and other agricultural produce. Other form of economic activity in the village was subsidiary to agricultural production and meant to provide the needs of the farms and of the farming population. The familiar characteristics of a capitalist society, namely wage labor and capitalist exploiter, were not the broad features of such a form of socio-economic organization.

How long such a society would have continued if the British rule had not supervened is difficult to say. There is evidence that capitalistic forms of organization had already made their appearance in the industrial and commercial fields. In the agricultural field also, the old village organization on the Western coast of India as also in Bengal had already been modified when the British appeared on the scene as the political authority.[39] But, speaking broadly, the socio-economic organization in the rural areas was still predominantly communal and corporate. The British rule provided the cataclysmic agency which forced the pace of change. The social cost of the economic change was borne by the vulnerable sections of the rural community like the agricultural labourers, the small tenant farmers, the village artisans and the menial servants in the village.[40]

Two factors were principally responsible for the revolutionary changes with profoundly affected the agrarian structure of the country (a) the destruction of indigenous industry and trade in the later of the eighteenth and the early nineteenth century and (b) the British land revenue system which created the proprietary rent receiving interest in land that came to be freely bought and sold. To these were added, later, the effects of the introduction of the railways which revolutionized not only the basis of agricultural production but also helped to bring in the village, the market relations and cash nexus

(a) Destruction of Indigenous Industries of India

Till about the India end of the eighteenth century, we are told by Vera Anstey "the economic condition of India was relatively advanced and Indian methods of production and of industrial and commercial organization could stand comparison with vogue in any other part of the world.[41] Infacts Inland had to resort to heavy tariff duties on Indian imports to protect her Woolen industry and prevent drain of gold to India. It was then that British begin to forge a head in industrial production with the help of new techniques and inventions. Indian industry depending as it did on excellence of crafts man ship rather than on steam and machinery, would have, in any case, suffered in competition with the British manufacture because it had failed to adopt the new techniques of production. But the establishment of the British political supremacy in this country coincided with the Industrial Revolution in Britain and the new rules of India did not hesitate to use their political power to smother the manufacturing industry of the subject nation. The export trade in India manufactures was destroyed by deliberate acts of discouragement and by imposition of heavy import duties on Indian manufactures in England.[42] This destructive process seems to have been completed by about 1833 in which year the East India Company lost its rights of trade and was converted primarily into an administrative agency. Meanwhile England had begun to push forward the exports of her manufactures of India. The value of British cotton exports to India shot up from £0.11 million in 1813 to £3.86 millions in 1840, £5.22 millions in 1850, and £6.30 millions in 1856.

In 1814, the quantity of cotton piece goods exported to India from Britain was 818,208 yards: in 1835 the figure had risen to 51,777,277 yards. Thus British manufactures had already made sufficient inroads into the Indian market before the railways began to traverse the length and breadth of the country. [43] After the revolution in means of transport, the flood of cheap imports from abroad not only undermined the competitive power of the urban weaving industry producing finer cotton and silk fabrics, but it also made its way into the rural areas thereby threatening the livelihood of the village weaver.

Though the most important, the cotton textile was not the only industry which suffered as the result of foreign competition Urban industries producing silk, woolen cloth, brass, copper and bell metal wares, enameled jewellery and stone carving, embroidery and manufacture of paper mache met a similar fate. Industrial labor was thrown out of its traditional occupations and in its search for livelihood turned to agriculture. The village artisan, except the weaver, did not have to face any direct competition from abroad. But the gradual change in tastes and modes of living of the people that came in the wake of the opening up of the village to the influence of foreign modes of living and consumption led to the gradual replacement of locally made products by chap machine-made substitutes. The potter, the oil expeller and the shoe-maker lost their custom in the village and were gradually reduced to the status of wage laborers.

A distinct class of landless day laborers merged in the rural areas by the middle of the nineteenth century and became more numerous in subsequent decades. Sir Thomas Munro, the Census Commissioner in 1842, considered the number of landless day laborers' employed in agriculture at the time to be too insignificant to require statistical measurement.[44] In 1852, George Campbell declared that in India "as a rule farming is not carried on by hired labor."[45] But 20 years later, 18 percent of the total agricultural population was counted in the census of 1872 as agricultural labors. This change in the structure of the rural society was of far reaching consequence. On the one hand, it freed labor from the shackles of the custom-ridden feudal society and gave it an independent status on the other, however, it exposed it to the uncertainties of the labor market due not only to the seasonal character of Indian agriculture but also because of the vagaries of rainfall.[46]

(b) Change in Land Tenures

While the decline of Indian industries and village crafts altered labor relations, a more fundamental change in the rural society take place as a result of the development of new property relations. Land in was, so long as the village communities were intact, a source of agricultural wealth only. With the distinction of the old rural social structure and the introduction of the new concept of landlordism modeled on the British pattern, agricultural land assumed the position of a capital asset which began to be acquired, from about the beginning of the nineteenth century, for the rent it yielded to products were needed mostly for farm work which continued to be carried on in the traditional way, but the construction activity in the urban areas offered them opportunities of employment which were readily availed of by them. Agricultural land had not acquired a market value in any part of India, expecting on the Malabar Coast, till the beginning of the nineteenth century.[47] The Revenue and survey settlements, the record of rights, the new notions of "property" and the Civil courts made it a freely transferable asset and gave it a market value. The urban capitalist class finding the older channels of investment being gradually narrowed down turned its attention to the acquisition of landed property by direct purchase or by advancing loans against the security of agricultural holdings. The competition for land gave it a new value. With the increase in trade in agricultural produce and the rise in agricultural prices in the latter half of the nineteenth century, land gained in value steadily. It becomes a valuable security against which the land owner could borrow easily. The credit of the agriculturist went up and with it raised the volume of agricultural debt. The high rates of interest charged by the village banker together with insecure and low yields in agriculture made repayments of loans, once contracted, almost an impossible task. The result was the dispossession of the cultivating classes and rapid transfer of land to the moneyed classes. The Deccan Riot Commission found in the records "relating to this subject from 1850 to 1858" two features "which had already become marked characteristics of the relations of sowkar and ryot under the altered conditions of our revenue and judicial system." These, the Commission observed, were "the growth of the small capitalist class engaged in money lending and the unequal operations of our laws to the disadvantage of the ryot." [48] in the North Western Provinces, Colonel Baird Smith found in 1861 that "sales of land for payment of public revenue had almost disappeared from the public records while sales by voluntary action or in satisfaction of debts had vastly increased."[49] He quoted figures from various districts indicating the loss of estates by agricultural castes in favour of the commercial classes. In Fatehpur, at the time of the cession of the district to the British,

The Mahajans (moneylenders) did not hold a single estate while in 1860 they had 86. Similarly in the district of Kanpur the Mahajan class was holding, in 1860, 301 estates and paying one-sixth or one-seventh of the total revenue of the district. For the province as a whole, there was a "fradual growth of moneyed classes and the progressive tendency of these classes to seek investments in land." Similar facts were brought out about the Punjab by Mr. Thorburn in his Musalmans and money lenders in the Punjab [50] and in the later enquiry into the subject conducted by him under the orders of Sir Dennis Fitzpatrick, the then Lieut. Governor of the province. An instance of considerable interest which throws some light on the gravity of this problem is that of the areas transferred from the Gwalior State to the district of Jhansi in 1886. As soon as the new lands were transferred to the British terror they became saleable property and thus acquired a value which they did not possess before. Thus, immediately after transfer, they began to pass from the possession of the agriculturist with such rapidity that the special Act under which the district had been temporally relieved had to be extended to the newly incorporated territories.[51] Throughout the British territory, excepting perhaps in the Madras Presidency,[52] there was thus, in the last 40 years of the nineteenth century, a strong trend towards the transfer of land from the agriculturist to the non-agriculturist money laundering the property laws introduced by the British, the increase in the value of land on account of the rise in the price of agricultural produce and increasing pressure of population on land, the administration of justice in the civil courts which were obsessed with the ideas of "property right" and "sanctity of contract."[53] And the agriculturists' demand for the credit to meet the requirements of working capital and other cash needs provided the money lending classes with an opportunity to enrich themselves by acquiring hold on agricultural lands and crops and they made full use of it.[54]

The new class relations introduced in agriculture affected both production and distribution. So far as production was concerned agricultural capital came to the separated from labour engaged in the cultivation of land. This retarded agricultural progress, because the cultivator lacked capital to invest in improvements. The moneyed classes, on the other hand, found money lending and acquisition of agricultural property more paying for the rent it yielded and did not, therefore, generally interest themselves in making improvements in land or in other agricultural equipment. As regards income distribution, a large slice of agricultural income came to be paid by the cultivator to his landlord and moneylender. Agricultural labour was no more able to enjoy under the new conditions the fruits of its own efforts. Rack-renting and exorbitant interest charges not only impoverished the cultivator but also destroyed all his incentive for agricultural improvement and progress. The situation was well summed up by the Calcutta Review which wrote: "The increase of wealth outside the (agricultural) community and the increase of poverty within disturb the unnatural relations which exist between the labour of the community and the wealth of the outsiders.[55] Adam Smith's theoretical mode[56] of the rent of the Landlord forming the first and the profits of the capitalist the second deduction from wages of labour when a society transforms itself from the rude and primitive stage into capitalistic organization, came close to reality in the context of the Indian situation. The deductions that interest and rent constituted proved so heavy that the peasant only managed to subsist in normal time on what was left after these fixed charges had been met. In the years of drought and crop failure, he had either to borrow against the security of land, if he had any, or go the relief houses.

Colonialism and Famines

The economic transition in India was bound to be accompanied by hardship for those who failed to keep pace with economic progress or take advantage of the opportunities that it offered. England had a similar experience in course of the Industrial Revolution. The poor Law of 1834 was a recognition not only of the existence of economic distress among destitute and paupers amidst all-round rapid economic progress, but also of the States responsibility for relief to the poor. But the magnitude of hardship was much greater in India in England, the changeover was rapid and people thrown out of employment in the rural areas were soon absorbed by the new industries. In India, on the other hand, the process of industrialization was retired by the compulsions of colonial economy. The entire population was, therefore, increasingly thrown on agriculture as the principal source of livelihood. Thus in India the normal economic process was reversed. In England and other European countries, in the nineteenth century, labour to the expanding industries, in India, manufacturing industry threw out labour to be absorbed in stagnating agriculture.

The burden which Indian agriculture was called upon to bear increased for another reason. The expenditure of the Government was on a lavish scale and quite out of proportion with the economic conditions of the people of the country.[57] Besides, the country had to bear the cost of Imperial wars in Burma and Afghanistan. The depreciation in the value of currency that began in 1873 and continued over the next twenty years increased not only the burden of Home Charges but also created the demand for compensation from the British officers in India. The principal sources of revenue at the time were land revenue, excise salt tax, stamps and opium leaving aside the opium revenue, the burden of all these taxes fell on the rural classes in particular. With the expanding expenditure of the State, the Government became increasingly exacting in its land revenue demand in the provinces in which permanent settlement had not been effected. The settlements were made at short intervals and every new settlement resulted in the enhancement of the Government demand. The policy of increase in taxation was opposed by the Madras Government as early as 1859.

The Governor, Sir Charles Trevelyan wrote four spirited minutes of protest against it.[58] Even the highest authority in the Indian administration of the day, the Secretary of State for India (Lord

Salisbury), confessed to the inequity of the Indian tax system and wrote in a famous minute in 1875 the following memorable words: "It is not in itself a thrifty policy to draw the mass of revenue from the rural districts where capital is scarce, sparing the towns where it is often redundant and runs to waste."[59] He went on to say that "the injury is exaggerated in the case of India, where so much revenue is exported without a direct equivalent" and added that " as India must be bled, the lancet should be directed to those parts where the bold is congested, or at least sufficient, not to those which are already feeble from the want of it.[60] Ten years later, the then Finance member of the Viceroy's Council expressed himself in somewhat similar terms. "One great blot on our administration not only still unremoved but aggravated by the course of events in recent years," said Sir Auckland Colvin,[61] "is that the classes in this country who derive the greatest security from the state are those who contributed the least towards it." In spite of the strong opinion on the subject held in such high quarters, circumstances prevented the Government from shifting the burden of taxation from land to the moneyed classes. The Government of India was precluded by superior orders from London to look to customs for any large addition to its revenue. Income tax was levied in 1886 but the tax receipts were small because it could reach neither the landlords who were exempted from it nor the moneylenders. Its rates could not be fixed high because of the opposition to the tax from the British merchants and civil servants in India who were the main parties affected. In the circumstances, agriculture had inevitably to bear the cost of the country's administration. Besides there were the interest payments of British investments in the railways which till the end of the nineteenth century remained a losing concern, interest charges on India's public debt held in England, and payment of salaries, furloughs and pensions of the civil and military personnel in the service of the Government of India. Whatever the justification or otherwise of these payments, it has to be Remembered that these could be made only by maintaining an export surplus of far as the balance of trade in merchandise was concerned. As the India economy had been transformed into the classic colonial pattern by 1860 and exports consisted mainly of agricultural produce including food grains. The Home Charges were met principally by the export of these commodities that is the reason why in spite of the increasing frequency of famines the exports of food grains showed a rapid and remarkable increase after 1870.

Price and Wages

While the mass of agricultural population suffered adversity due to various circumstances, analyzed in the foregoing pages, the agricultural labour was particularly hard hit by the adverse shift in the relation of wages to food prices during the period under survey. There was a secular rising trend in prices of food grains over the whole period while the wages of the agricultural laborer in the interior of the country remained almost stationary till about the close of the century. India experienced for the last forty years of the nineteenth century, with some interruption, a book in the grain market and a continuous depression in the labour market.

We shall not enter at this stage into a discussion of the causes of the rise in food prices during the period under study, but that prohibition of exports' of food grains[62] would have kept down food prices will not be denied by anybody.[63] Food grains became after 1883 the largest single item in India's export trade and remained so till about 1914. There was nothing else to take their place if their exports were prohibited. The possibility of any reduction in the total value of exports was unacceptable both to the Government and to the British economists. In the words of Atkinson, "to adopt any measure that would tend to reduce exports would adversely affect the balance of trade of the country, and this, if continued would adversely affect the Gold standard (after 1893) and the rate of exchange, and might cause of reversion to the silver standard, with the accompanying heavy home Charges caused by a fall in exchange." In addition, Mr. Atkinson continued, "if the food supply from India ceased, unless the gaps could at once be filled from elsewhere, food prices outside India would rise, and this owing to the existence of unions and in opposition to economic laws on the subject, might affect wages outside India and thus indirectly all prices. Obviously Imperial considerations were of far greater importance to the Government of the day than the interests of the people of India. Thus, although the exports of food grains and a rise in their prices brought prosperity to certain sections of the populations, in the absence of a commensurate rise in wages, high prices were a source of persistent hardship for the laboring classes.

For the normal situation in agriculture seems to have been, even at the beginning of the period under study, one of under-employment.[64] Wages in agriculture could, therefore, be raised only by expanding employment opportunities outside agriculture at a rate faster than the natural increase in population. But the actual situation in the country over the last forty years of the nineteenth century turned out to be quite the reverse, the economic development of the country was slower than the increase in population. The problem of under-employment in agriculture was, therefore, not only not solved but steadily grew worse until the beginning of the present century when partly because of the heavy mortality caused by famines and plague, and partly because of the increased employment opportunities Offered in mining, docks, railways, commercial, houses, and textile industries, wages begin to rise even in the rural areas.

It is against this back ground of poverty of the agricultural masses and growing destitution among them, rising prices of food grains and stationary wages, and the increasing population at the scope for employment outside agriculture that a series of famines occurred in the country. Violent fluctuations in income employment, prices and wages which a famine implies should have produced considerable distress even in a rich country.[65] Their effect on the poverty-stricken masses of India was bound to be calamitous, it is with this story of the frequent convulsion from which the Indian economy suffered during the period of its transition from the old to the new order that we are concerned in this work. We shall examine the economic and social changes and their effect on the economic condition and the standard of living of the agricultural masses, the famines and scarcities that occurred and the economic policies that the Government adopted and the administrative action that it took to relive distress and prevent the future recurrence of famines. Chapter XI is devoted exclusively to the Bengal Famine of 1943 while the concluding chapter discusses the food problem in the context of rapid economic development in India in the post-Independence period.

Drought and Agricultural Production

The united provinces of Agra and Oudh experienced one of the worst famines in history in 1907-08, caused a decline in the out-turn of Kharif crops estimated at thirty one percent of the normal out-turn as against thirty seven percent. The total loss of crops in the autumn of 1907 was an estimated four million tones. This meant a loss of food for 48 million inhabitants of the province (Bhatia 1991: 151). The occurrence of droughts has continued into the current times. The drought related crop failures of 1967 and 1969 had the potential to bring about the worst famine in the sub-continent's history. There was a nineteen percent fall in grain production in 1966-67 from the previous crop season. Grain stocks were used up, leaving few reserves when drought and poor crops recurred in the following year (Hewitt 1983: 190). These were the years of import of very huge quantities of food grains.

During the early 1970s Maharashtra suffered three successive years of drought which culminated in disastrous failure of the monsoon rains of 1972 and the associated calamitous agricultural year of 1972-73 when cereal production fell to under half of 1967-68 level. Even in good agricultural year Maharashtra was a cereal deficit state per capita cereal production declined to barely 51 kg per person, only 47 percent of the level of 1967-68 and only 71 percent of what might have been expected even on the basis of the declining trend of the previous four years. Turning to the districts which lie at least partly in the drought prone zone, a more complicated picture emerges. Cereal production per capita across the districts in 1972-73 was barely 45 kg, only 37 percent of the level of 1967-68, and well below the trend. More over all of what were termed the "most acutely affected" districts were located in this zone (Dyson 1992: 1325). The state of Maharashtra (with an area of 3,08,000 km and a 1971 population of 50.4 million) experienced a three year drought in 1970-73 during which the output of food grains (cereals and pulses) were 19 percent below the average of 1968-69 and 1969-70 in the first year and 53 percent in the third year. These deficits were added to Maharashtra's normal need to import 2.6 million tons of food grains, about one third of its consumption requirements (Me Alpin 1987: 394).

The production of food grains in 1972-73 was below that of 1968-69 by as much as 54 percent, reduction in production of food grains was to an extent of about 80 percent in Bihar, Solapur and Sangli districts between 65 and 75 percent in Aurangabad, Nasik, Ahmednagar, Nanded districts between 55 and 65 percent in Satara, Dhulia Osmanabad, and Akola districts and the reduction was around 50 percent in Pune Thana, Parbhan, Buldhana and Bhandra districts (Brahme 1983: 51). During the 1972 drought (1972-73) there was a reduction by about twelve percent in the total area sown in Maharashtra. A decrease in the hector age to the extent of about 20 percent was noted in the case of pulses and chillies and thirty percent in the case of bajra. The corresponding percentages were 15 for wheat and 12 for groundnut. The extent of reduction in the case of crops like Jowar and cotton, was less than ten percent while there was no decrease in the area under rice and sugar cane (Brahme 1983: 51). The impact of drought on the village economy can be identified by considering variables such as the area under crops, and the production of various crops which undergo deceleration. Drought prone areas in Rajasthan show the occurrence and impact on local crop yields in different years. During the eleven year drought hit period the agricultural productivity in as many as seven years recorded a crop loss of 95 percent (Joshi 1983: 45).

Livestock as During Droughts

Among the assets, decline will be much in drought animals followed by the livestock as during droughts, animals perish or sold out, in grain prices rise rapidly, while animal prices fall and the herder's purchasing power collapses. The high incidences of selling and mortgaging of land and livestock during famines show the magnitude of distress suffered by the farming community. Singh (1965) noticed that such distress sales were prevalent particularly among the middle class and small group of farmers (Curry 1976: 71). During the 1972-73 drought in Maharashtra, the loss of drought animals varied from about 20 percent to 35 percent in different villages. In the face of acute scarcity of water and fodder resources it was difficult to maintain the livestock, particularly the bovine population in the drought affected villages. Loss of livestock was estimated to be 21.8 lakh which comes to 13 percent livestock of 14 districts in the state. It was estimated that about 6.1 lakh of cattle were lost due to death and 8.2 lakh of cattle were sold. The total loss inflicted on the cultivator on account of the loss of livestock was estimated to be of the magnitude of about Rs.61 crore (Brahme 1983: 51). During drought, fodder as well as food grain become scarce. Drought may, in certain situation result in increasing deterioration of land and pastoral resources exert pressure on productive lands and the development of a land market characterised by speculation and absence of guarantees in transactions. It may also result in appropriation of animal capital, pastoral land and of watering points by the influential people. Cattle perished due to acute scarcity of fodder and drinking water during the 1979-80 period (Madhya Pradesh). In Maharashtra, during the 1992 drought as many as 1,292 villages (of Gadchiroli district) faced unprecedented shortage of food and fodder for cattle (Dhangere 1992: 1421).

Drought and Indebtedness

According to the Famine Commission of 1890, in times of very great scarcity, prices of food grains rise to three times their ordinary level. "Whereas in ordinary years the prices of food grains of the mass of the people may be from 20 to 30 seers a rupee, in times of great scarcity, it will rise to 8 to 10 seers a rupee or even higher. During the 1907 drought in Bundelkhand, prices rose sharply wheat touched the peak of 6 seers a rupee on 15th October 1896" (Bhatia 1991: 151). The famine of 1837-38 in Uttar Pradesh covered an area of about 25,000 square miles which was inhabited by a population of about 8 million. Drought was accompanied by scarcities and high prices which rendered the lower classes to great extremities as in the Fatehpur district. The prices of food grains rose high casting much distress amongst the lower classes (Allahabad). Some believed that even the price rise was not of great consequence as the people lacked the ability to purchase food (Sharma 1993: 341)

An enquiry by Thorburn (Bhatia 1991: 52) in four selected circles of the Punjab revealed not only the existence of a state of widespread indebtedness among agriculturists but also a rapid transfer of property from them as a result of that. In 474 villages, in widely distant and differing tracts held by different tribes and colonies, the cultivators in 26 villages were helplessly involved in debt, in another 210 villages they were seriously involved in debt, and only in 138 villages the debt burden was comparatively lesser.

The famine of 1858-1900 in Rajasthan (Rajputana) resulted in extreme social backwardness of peasantry who had relied for centuries on the feudal landlords. On account of a backlog of indebted amount the peasants were forced to pass on the land into the hands of non-working classes leaving the tenants and the landless labourers in misery. Increased indebtedness resulted in lack of capital for agricultural works and there was also absence of alternative outlets for employment which forced them to the margins of abject poverty. Indigenous craftsmen, specially the weavers were impoverished as drought restricted their markets (Maloo 1987: 234).

During the 1979-80 drought in Madhya Pradesh the incidence of poverty was found to be higher (68 percent) and for SCs and STs it was 75 percent. This sort of high level poverty among sampled households was attributed to the low returns from agriculture and also partly to the methods of cultivation (Balakrishna 1982: 27). Drought is normally characterised by water shortages. The 1979-80 drought in Madhya Pradesh when 23 districts were declared as drought prone, Drinking water was a major problem in most of the villages. During the 1992 drought in Maharashtra as many as 1292 villages (located in Gadchiroli) suffered severe drinking water problem. Many of these villages either do not have any wells or, if they have, the wells have dried up. Rural peasants, tribal households and women were in desperation as they have to meet their needs of drinking water by resorting to various alternatives. In Maharashtra (Vidarbha and Marathwada) out of over 42,000 villages as many as 29,157 (nearly 22,000) of the total (29,157) villages declared as affected by water scarcity and faced famine like conditions that were nothing short of catastrophe (Dhangere 1992: 1421).

Data covering 19 villages from 7 states (Gujarat, Karnataka, Kerala, Rajasthan, Uttar Pradesh and West Bengal) revealed (Benerji 1982: 105) a substantial group of rural population, comprising small farmers and artisans who manage to get two full meals all the year round, live in almost perpetual fear of being pushed down into the "hunger classes" by a variety of natural calamities and individual catastrophes such as prolonged sickness and death of the breadwinner, old age and crushing social obligations like marriage expenses.

Diseases literally thrive under the conditions that are prevailing in the Indian villages, and a vast majority have to fight these adverse conditions right from the birth. The mother is exposed to serious hazards which cause high mortality and morbidity rates. Their children face acute problems of nutrition and infection in the very early phase of their lives. During the great Bengal famine of 1943, devastating epidemics of malaria, small pox and cholera were associated with the famine. Famine diarrhea, famine driopsy and anemia also took a very heavy toll of lives (Benerji 1982: 105).

Short of starvation deaths, every kind of damage occurred to an alarming degree of hunger and sever nutritional deterioration during the 1966-67 drought in Bihar and food crisis based on large scale scarcity during 1970-73 drought in Maharashtra have been reported. During 1992 drought (Gadchiroli in Maharashtra) consumption of pollutant and undrinkable water has given rise to epidemics of cholera and gastroenteritis in the district (Dhangere 1992: 1421).

Drought and Socio-Economic Structure

The operational impact of drought would explain the twin processes of marginalisation and stratification of societal groups as these groups are often disturbed by drought. Historical evidences provide scope for generalisation of famine induced droughts which resulted in hunger and malnutrition. Droughts in the present context create serious problems for men, women, children, livestock and their environment.

Drought's impact can be identified in location specific semi-arid areas where the rainfall is low and the irrigation is of scarce. The primary concern of the present study is the impact of drought and responses to it in bad years, the socio-economic impact of climatic fluctuations favors the paradigm of human ecology which characterises households as actors in their environment adjusting and adapting to it's changes. By contrast, during a drought episode, households balance their resources and objectives by adjustments, which are short-term coping strategies. Since the farmer's decisions would (largely) determine the effectiveness of various measures adapted, it is necessary to examine the farmer's adaptation to rainfall variability. Impact on crops, production and livestock loss also form the basic criterion to determine the effect of drought in a region. Within this socio-economic framework, the present study makes an attempt to study the cropping pattern, asset mobilisation, migration and livestock movement, which are the only alternative strategies during drought.

Drought affects the agricultural sector where farm level activities are disturbed due to lack of rains. Water deficiency causes food grain deficiency, which in turn leads to increase in price levels crop losses results in seasonal unemployment and losses in agro-industries and regional income reduction. Employment opportunities become scarce during drought in the area to causal labourers, daily wage earners, marginal farmers and agricultural labourers suffer following risks in subsistence agriculture. Low agricultural yield results in low income, shift in living standard variation in the economy, fall in regional output and inflation becomes incredible. Poor nutrition due to food shortages to children, malnourishment to agro-pastoralists and hunger becomes common during scarcity period.

Drinking water shortages due to minimal groundwater recharge and water contamination (in fluoride areas) occur as intensity of drought increases. Rural women, children and tribal people suffer due to the distance they have to walk for water, cattle also suffer due to water shortage. Scanty rains leading to prolonged depletion of pastures result in either non-availability or limited availability of fodder and water for cattle. Meager food, water and fodder supplies result in competition between human and livestock. This would in turn cause sale of livestock such as cows, bullocks, sheep and goat at low prices which in the long-run decrease farm efficiency. Migration and transhumance to distant areas isolate rural households and cattle from local living. Fall in dairy production also can be seen as the milch animals are either subjected to sale or migration. Drought also causes a decline in livestock products and their quality while following diminishing livestock assets.

Drought simply destroys or disturbs the ecosystem through soil and land degradation process. High population density on limited resource zones, land use problems related specifically to grazing such as conflict between local and neighboring pastoralists for water and pasture, desertification and degradation, fragile lands become common in the event of a prolonged drought. Soil deterioration takes place due to continuous monocrop cultivation without adding much to soil fertility, soil loss by over-cultivation, grazing and indiscriminate wood cutting resulting in an increase in soil erosion and run-off.

The other visible effects of human action in drought affected pockets, in addition to soil cover depletion and denudation of tree cover and species in the long run, are low quality of pasture and tree crops, sand dune formations in wind locations. An increase in cost of water use and water management can be observed in certain areas where groundwater would be available only at 200 to 250 feet depth. This necessitates huge investments in digging bore wells and overhead tanks. In these areas, water quality gets affected as fluoridation systems are yet to be made cost effective.

Another measure to reduce the drought effect is to go for soil conservation through deep ploughing, leveling, bunding, deep terracing and furrow method. Conservation of water resources along with soil becomes essential. Millet farming, sheep and goat rearing and watershed and grazing residue management through fodder crops and plantations fetch economic benefits. Further flood protection, waste land reclamation, irrigation systems, agro forestry schemes are needed to mitigate rural drought. Such measures of drought relief are essential features of a short-term coping strategy. During drought it is the agricultural and non-agricultural labourers, landless poor, small and marginal farmers who become vulnerable to credit as they move to urban areas in search of work to earn income for daily maintenance in the form of food grain and wages. Non-availability of employment opportunities within the local area and lack of food makes essential the house holders either to participate in public works or migrate. Scarce the irrigation to reel under poverty, also increase pressure on common property resources such as villages commons, pasture, grazing and forest.

Following rainfall variability and scarcity farmers are compelled to change the cropping pattern in favor of dry, less valued, crops. Usually changes occur in food crops from paddy and pulses to dry crops such as jowar, bajra, cotton and groundnut. A decline in cash crops results in subsistence farming with a decline in cropped area and on-farm employment. Since there is a lack of insurance against bad weather, possibility for changes in input use is limited during severe drought.

Migration becomes one of the foremost adjustment strategies. Peasants with small and marginal holdings of land migrate to irrigated areas to harvest paddy or to urban areas to work, persons with enough resources may resort to self-employment. A majority of rural labourers are left in seasonal unemployment and would like to participate in government and public works. Changes tend to occur in food consumption pattern as households either retrench or shift their diet from consuming rice and dal to nourishing food prepared out of jowar or bajra. Past savings are exhausted and borrowing of money and food grains, selling or mortgaging of assets, postponement of dues appear in general. To tackle drought, farmers follow adjustment strategies such as subsistence production, inter-cropping, crop combination and phasing, better use of water and land resources, shift from staple food crops of favorable years, agricultural diversification, seed and fertilizer combinations. However all these modernization techniques are either availed by resourceful farmers or only limited to irrigated zones. Government can intervene early to avoid havoc through employment generation works by assuring minimum wages. Stabilization of food grain production and food availability, farm infrastructure, farm prices and cattle prices can offer much to the poor farmers in times of helplessness than providing subsidies, which are often reaped by the well to do state of the peasant community. Improving food delivery system through proper functioning of public distribution system to areas which are severely hit by the drought will enhance the relief measures. Infact it is one of the focused areas of relief operations. Essential commodities are to be supplied at reasonable prices to the low-income rural households. Clean drinking water establishments, fodder provision through setting up of cattle camps, proper credit sanctioning in the form of short-term loans with low interest rates and reasonable subsidies are the other measures. Local level community organizations can help in proper utilization of common property resources. Drought loss management through supplementary earning, sustenance relief and post-drought recovery are essential. Drought affects different socio-economic strata of the rural society differently. Although drought affects the entire rural economy adversely, it is the small, marginal farmers, agricultural labourers and other poorer sections who suffered most. Analysis of loss in per acre yield of Kharif crops by the size of holdings show that small cultivators often suffer a lot than the medium sized and big cultivators. It is the small cultivators, medium sized cultivators and the big cultivators who incur money loss due to drought.

The average monetary loss incurred due to drought increases with increase in the size of holdings. The maximum reduction is in the production of big farmers who suffer when compared with small and medium cultivators. The pasture fields affect the grazing lands in case of small cultivators. The impact on total incomes of the three categories of farmers follows another trend.

The small and medium sized farmers suffer less when compared to big farmers. The unemployment problem show a reverse trend as pervasive in small cultivator's group and none from the categories of medium sized farmers or big cultivators perceive any problem of unemployment or changed its occupation as a result of drought. Dearth of food grains show a smooth trend to move up with the size of holdings, probably due to the reasons of either the bigger cultivators maintain a bigger size of families, joint families for they estimate shortage after adding food grains required for hired labour in their normal household requirements. The impact of drought is detrimental especially for the small and marginal farmers (AERC 1972: 10). In the two villages of Mahubhata and Sripalli (Kalahandi) there were two groups namely Kumbars and Kurmis. Kumbars are traditional potters and Kurmis belong to and agricultural caste and most of them depend entirely on agriculture. Due to land concentration in a few hands, and drought for three years, the poor belonging to the aforementioned sections suffered more as they were compelled to migrate to other areas in the absence of livelihood. At the same time the big and rich farmers could get rice and wheat throughout the year. The middle-class showed a favorable tendency towards consumption, access to education and also to the schemes undertaken by the government.

The poor conversely, suffered due to crop loss and weakened economic base. The small and marginal farmers, and labourers who depend on agriculture were largely affected by drought. The middle class by virtue of its structurally advanced economic position could provide for an ideological rationale of developmentalist's process as operating in India today (Nayak and Mahajan 1991: 58). The famine effects are neither random nor even on various peasant sections.

Population loss was more among Madigas, Boyas and Chenchus (Kurnool). There was a change in the age and sex composition of families belonging to small peasants of depressed castes. The great Bengal famine, particularly during 1942 to 1945 caused increased economic differentiation which is of interest. The landlords and professional class were unaffected but there was a net downward shift of households from middle economic status (ryot, small artisan and trader, petty employer, rentier) to poor economic status (ryot-bargadar labourer), while a large number of households belonging to the poor economic status classes (ryot-bargadar labourer) become destitute, and either emigrated from the village altogether or were reduced to beggary. The slow disintegration of the middle group of self employed peasants and artisans, which had been going on for decades was found to have been accelerated. Further there has been acceleration in economic differentiation within the peasantry, under the impact of war and famine conditions (Patnaik 1987: 11). The people living in a village called Ranawatan-ki-sadri in Chittorgarh of Rajasthan provide another example with land and labour as primary factors of production of a wealthy system of land tenure and the feudal forces at work a rural system. The landlord (Jamindar) and his relatives seized the power and resources whereas the people belonging to lower castes suffered due to their caste oriented activities. Rainfed crops suffered due to the absence of canals for getting water from tanks, while thousands of people died in famines. However sheep and goats helped some people in meeting the domestic necessities, and in fact formed a source of subsidiary income to the farmers (Chauhan 1967: 12).

Any solution to the serious situations of drought has to begin with the analysis of its impact on the weaker sections who suffer the most. The basic problems that have led to drought and resulted in the marginalisation of the people have to be dealt with it is not only tribals but also forest dwellers who suffer more due to deforestation. As far as the small person is concerned the environmental and social problems have to be dealt with (Fernandez 1987: 421). Environment has to be viewed primarily as people who have a symbiotic relationship with it. Lastly, the Balutedari system in Maharashtra, Mirasi system in South India, are contemporary to the Jajmani system found in northern India. Jajmani system helped the rural artisans as well as the local landlords to survive through the difficult phases of drought. The Jajmani system proved a kind of inherent survival strategy to the artisans particularly to the washermen and carpenters, during the drought. When the drought aggravated in rural villages, (for example survey villages in Telangana and Rayalaseema in the present study) consisting of small and marginal farmers and agricultural labourers migrated to other areas to work. But those (particularly washermen, carpenters) working under jajmani system with big farmers and landlords could survive the drought without migration, through serving them with works such as washing cloths and making agricultural tools and implements such as ploughs and bullock carts. Inspite of crop failures the lend owners paid essentials to these artisans while adjusting the some with other non-farm incomes. Thus the jajmani system is still prevalent in a number of villages in south India which not only represents the traditional bondage between the landowners and the servants but also reveals the underlying drought survival strategy in rural India.

Food Security during Famines

In order to review the history of twentieth century famines, it is useful to go along the footsteps of those social scientists dealing with this phenomenon. Devereux notes that most definitions characterize famine as a discrete event that is triggered by food shortage and results in mass death by starvation.[66] Famine theories evolve out a long period of intellectual processing under deferent paradigms which influenced time by time proposed analysis and policies. In an historical perspective, a major role has been played by influential views of Thomas Malthus, even if in 1980s a series of studies challenged some of his assumptions. It is the case of Amartya Sen[67], which elaborated an alternative approach to famine that described food crises as failures of "entitlements". In so doing, he shifted the analysis from an exclusive focus on food shortages to an emphasis on people's inability to access food, whether or not a shortage was a contributing factor. In a series of articles critiquing Sen's approach, Rangasami went further, arguing that "famine should be viewed as a process with distinct phases, rather than an isolated, aberrant event"[68]. In another work, Alex de Waal contested the definitional association between famine and mass death by starvation.[69]

While these studies challenged western conceptualizations of famine, there was an increasing recognition of the role of conflicts in newly emerging crises in the 1990s, for the first time there has been the clear perception that the very nature of famine had changed, from essentially "natural" to pre-dominantly "political" phenomena. A number of studies of the newly identified "complex political emergencies" (CPEs), emphasized the socio-political dimensions of famine, demonstrating how some groups actually garnered benefits from the creation and perpetuation of famine conditions (Edkins 2000). Devereux argues that two significant shifts characterised twentieth century famines over famines of previous eras. First, in terms of causality they were immensely more complex than ever before, when famines were interpreted as natural disasters, being superseded by complex negative synergies between natural triggers and political culpability. Second, during the latter part of the century, food crises became concentrated in sub-Saharan Africa, where interactions between drought and civil war, in particular, became the dominant causal trigger of famine.[70]

Moreover, it should be noted that the persistence of hunger in many countries in the contemporary world is related not merely to a general lack of aáuence, but also to substantial inequalities within the society. In fact, the dependence of one group's ability to command food on its relative position and comparative economic power other groups can be especially important in a market economy. Furthermore, recent times have witnessed not only a rapid expansion of market exchange, but also significant developments in the conditions of ex-change with nature, i.e. production. On the one hand, advances in agricultural technology have increased the potential for improving living conditions in rural areas. On the other hand, environmental degradation poses a grave threat to the livelihood of the rural population in many countries.

Famine in Rayalaseema

The ominous spectre of famine which haunted the three Rayalaseema districts of Cuddapah, Kurnool and Anantapur and the Telangana districts of Warangal, Khammam and Nalgonda has receded, though it has not vanished yet, due to rains during the last few days. The relief brought by these rains has, however, come a little too late, because the drought has already worked havoc with the standing crops. Extensive relief measures were undertaken by the State Government to relieve distress, and the Central Government readily responded to its call for financial help as well as for larger allocations of food grains. The immediate relief provided to the affected areas consisted of liberal grants in repair existing wells and dig new wells to repair and renovate village tanks and oilier sources of irrigation and water supply, and sanction of taccari loans on easy terms to needy agriculturists.

Food Prices Spurt

The drought led to a sudden and steep rise in the prices of food grains in the affected districts. It seemed as though overnight rice and millets, the latter being the staple cereal in Rayalaseema had disappeared from shops. The most urgent need was to provide food grains. In low income groups, particularly agricultural labourers. The Centre agreed to release 50,000 tons of Burma rice for distribution through fair price shops. It also assured supply of almost unlimited quantities of wheat and millets from imported slocks at controlled prices. Orders were issued to open 2,000 fair price shops, about 1,500 in Rayalaseema and 500 in Telangana. Even before the late rains revived hope, more than 50 percent of these shops had begun to function.

Now that the situation has cased somewhat, it should be possible to review the food position in Andhra Pradesh from the long range point of view. What happened in Rayalaseema this year is nothing new. The area has always suffered from the vicissitudes of the monsoon and consequent failure of crops. Scarcity conditions emerge in an acute form once every two or three years and are met by ad hoc relief measures.

A few hundred local wells are repaired, a few minor irrigation works are hastily desilted and a few road construction and other local public works are undertaken in an uncoordinated manner. Appeals for help go forth and fair price shops are opened and supplies rushed from somewhere. There has always been a distinct note of unplanned haste in the provision of relief.

Small-scale Industries and Minor Irrigation Works Needed

What is required is a plan for systematic development of the area through a network of small scale and cottage industries, for which the area has abundant raw materials. One of these is groundnut oil crushing and another is sheep-rearing and blanket manufacturing. The plan should also include permanent remedies against failure of rains through the implementation of an extensive well digging programme, the execution of the long-deferred Tungabhadra High Level Canal project, which would benefit Cuddapah and Anantapur districts and the supply of water for growing a second and short crop of millets from the Tungabhadra low Level Canal. A number of minor irrigation works like tanks should be taken up to store water during good years for use in periods of drought.

These anti-famine measures would require, as the Union Deputy Minister of Food, Shri M.V. Krishnappa, said an allocation of about Rs 50 crores, for which specific provision could be made in the Third Five Year Plan of the State. This proposal deserves to be pursued with vigour by the State Government, which should impress on the Planning Commission that a substantial allocation of this nature would effectively prevent recurring drought and famine in Rayalaseema, and obviate ad-hoc grants which are frittered a way in immediate relief measures.[71]

Why food Stocks Disappear?

A distressing development that immediately follows the expectation of a monsoon failure is the disappearance of food grains and other necessities from the rural areas. This has happened in the past and has happened again now, not only in the genuinely drought affected districts of Rayalaseema and Telangana but also in the highly surplus coastal districts of East and West Godavari and Nellore, One reason for this phenomenon is that huge stocks of paddy and rice held by millers and other purchasing agents in the surplus districts are stored away in godowns or disappear into the black market as soon as a crop failure is expected. The second, and an even more important, reason is that there is an immediate stepping up of rice purchases in Andhra by agents and merchants from Madras and Kerala which, with Andhra Pradesh, constitute the Southern Food Zone, within which movement of rice is free.[72]

This is what has been happening during the last few weeks, especially since1 famine conditions appeared in Rayalaseema and Telangana. It is stated that between 2,000 to 3,000 bags are daily moving out of rice markets in the surplus coastal districts to Madras and Kerala. Apart from raising local food prices, these large-scale purchases have caused acute shortage of rice in the producing districts themselves. A State which is more than self-sufficient in food has, thus, to solicit and secure large allocations of imported Burma rice from the Centre to relieve scarcity conditions within its own boundaries![73]

Distribution of Andhra Surplus: Minimum Requirements Only

The absence of any control on purchase and movement of food grains from Andhra Pradesh to Kerala and, to a lesser degree, to Madras is resulting in an unenviable situation for the people of Andhra and also the Government, which is being accused of pusillanimous acquiescence to an arrangement which brings distress to the local population. So long as the existing zonal arrangements and food distribution policies continue, this distress cannot, indeed, be avoided. Hut a remedy which had been suggested more than once in the recent past is that the minimum food grain requirements particularly of rice of Kerala and Madras should be ascertained in advance, and supplies of that much quantity of rice should he arranged on a Government-to-Government basis. This would stop hectic unrestricted purchases by merchants from the southern States in the Andhra markets.[74]

The formulation of an alb India food policy, reconciling the interests of different States and areas with the interests of the entire country and ensuring equitable treatment to surplus, marginal and deficit States has so far proved an intractable problem. With goodwill and imagination, it should, nevertheless, be possible in find a solution, so far as the Southern Zone is concerned, if the Centre accepts one of the two or three alternative propositions propounded by the Andhra Government during the last three years or so for procurement and distribution of its surplus rice.[75]

Dissension within Cabinet The political crisis in Andhra Pradesh, arising out of sharp and seemingly irreconcilable personal rivalries within the State Congress party, does not show any indications of being resolved. The feeble hope raised by the Union Home Minister's efforts to bring about a rapprochement is clouded by vehement statements issued by leaders of the group against one another. It is, however, clear that though the Congress High Command may not be able to impose a settlement, it will not permit any move aimed at displacement of the present Ministry.[76] But can it stick to this resolve, if it finds the dissidents strong enough to force the issue?

In the meantime we have the sorry spectacle of one Minister of the State Cabinet openly threatening to sponsor a censure motion in the Legislature Congress party against another Minister, and of the Chief Minister asserting that if such a motion was moved, he would core sider it a censure motion against himself. The question is granting that the present ministry somehow survives the present crisis and remains in office till the next general election. Will it be a united and homogeneous team a qualification essential for efficient and effective Government ?

Table 2: Compound growth rates of principal crops in respect of area, yield, and production in Andhra Pradesh for the two periods (Area in lakh hectares)

	Before Globalization (1981-1994)			After Globalization (1995-2010)		
	Area in lakhs hectares	Yield Kg/ha	Production (in lakhs hectares)	Area (in lakhs hectares)	Yield Kg/ha	Production in lakhs hectares
Rice	0.55	1.93	2.00	-0.1	2.70	1.9
Jowar	0.64	1.47	-4.93	-7.56	5.81	-1.77
Cereals	-0.33	3.54	0.35	-0.38	2.50	2.98
Foodgrains	-0.3	2.89	-0.59	0.3	1.88	3.26
Sugarcane	1.55	-0.15	1.41	1.07	0.53	1.45
Groundnut	1.28	2.01	4.48	-12.48	-2.03	-3.17

Source: Statistical abstracts of various issues. Ministry of Agriculture Government of Andhra Pradesh

Table 2 provides information on compound growth rates of principal crops in respect of area, yield, and production in Andhra Pradesh for the two periods yield growth rates of principal crops of Andhra Pradesh were at higher level than the production growth rates. Highest yield growth rates were seen for crops like jowar, rice, cereals and food grains in period-II. In the same period in respect of area variable of groundnut of jowar had negative growth rates. But in the period-I area variables had marginal growth for sugarcane (1.55 percent) and groundnut (1.28 percent). In the period-II food-grains had positive growth rates (3.26 percent) in production variable, followed by cereals (2.98 percent). On the contrary, groundnut had negative growth rate (-3.17 percent) in production variable, followed by jowar (-1.77 percent).

Table 3: Compound growth rates of area, production and yield of principle crops in Rayalaseema for the period (1997 to 2012)

	Globalization (1997-98 to 2012)		
	Area (in hectares)	Yield(Kg/ha)	Production (in tonnes)
Rice	0.47	1.36	1.95
Groundnut	-0.69	0.22	-2.71
Sugarcane	-3.37	0.19	-3.5

Source: Ministry of Agriculture government of India

Table 3 provides information on compound growth rates of principal crops in percentages in respect of area yield and production variables for the period 1997 to 2012. For rice crops in Rayalaseema production variable registered a positive growth rate (1.95) percent. But yield growth rate was only 1.63 percent. The growth rate of area was less than 0.5 percent, Sugarcane and groundnut crops registered negative growth rates in production and area variables. Highest negative growth rates were seen in case of sugarcane -3.9 percent. However, yield growth rates were slightly positive for groundnut and sugarcane in Rayalaseema region.

Conclusions and Suggestions

By and large the higher growth rates of principal crops achieved in the Telangana region during the different time periods were due to favorable agro-climatic conditions, better irrigation facilities, adoption of recommended package of practices of crops, and various government policies and programmes introduced in the country. The Costal Andhra Region acquired second position in agricultural growth, followed by Rayalaseema region. There is a large potential in all the regions of the State to increase yield and production growth rates of principal crops in the State as well as India. All the problems of the farmers such as getting remunerative prices for their agricultural products should be solved by taking appropriate policy, programmes and measures in the areas of production and marketing of crops and live-stock products. The Indian seed programmes for encouraging the development of new varieties of crops and protecting rights of farmers and plant breeders should be strengthened. The funds for agricultural research should be increased. Facilities for production and distribution of quality seeds at reasonable prices to the farmers should be ensured. The Integrated Nutrient Management Programme (INMP) should aim at supplying the potassic and phosphatic fertilizers to the farmers at reasonable prices along with increased subsidy. Rainfed Area Development Programme (RADP) should help small and marginal farmers and improve farming systems in the country. The National Mission for Sustainable Agriculture (NMSA) should solve the problems and issues related to sustainable agriculture in context of risks (production risks and price risks) associated with the changes in the agro-climatic conditions and pollution hazards. The National Food Security Mission (NFSM) launched in the county to enhance the production of rice, wheat and pulses should be strengthened in the country as well as States.

Finally, the state has got an important role in combating world hunger. Anyway, we should remember that "hunger is a common predicament, but this does not indicate the existence of one shared cause". In order to come to grips with the problem of hunger in the modern world, it is necessary to get a clear understanding of the different issues that constitute it. The distinction between the problem of food security and that of famine is particularly important.

Notes and References

1. G. Blix, Y. Hofvander and B. Vahlquist (eds.), *Famines A Symposium Dealing with Nutrition and Relief Operations in Times of Disasters*, Almquist and Wikselle, Uppsalel, Sweden, 1971.
2. Southward, 'Famine', *Encyclopaedia of Social Sciences*, vol. VI, p. 85.
3. Walford, 'Famines of the World: Past and Present', *Journal of Statistical Society*, vol. XLI (1878), pp. 433-526 for an authoritative list of famines that are known to have occurred in different parts of the world from 2000 BCE to 1876 CE.
4. One of the earliest records of famine is the Stele of Famine which was found over a granite tomb at the first Cataract of

the Nile. This gravestone is inscribed with the report of a famine that is said to have occurred 2000 years before the time of Abraham. See 'Fearful Famines of the Past', *National Geographical Magazine*, July 1917.

5. Approximately four hundred and fifty famines are known to have occurred in Europe over a period of 850 years from 1000 to 1850 CE. Vide Southard, *op. cit.*

6. According to John Boyd Orr almost two-thirds of the world population lives in a state of chronic hunger (vide Foreword to Josue de Castro, *Geography of Hunger*, p. 6). F.A.O. has estimated (Monthly Bulletin, January 1952) that 60 percent of the world's population gets less than 2,000 calories of food per day, which quantity is regarded the minimum necessary requirement for the health and efficiency of a person.

7. For an informative account of social causes of Indian Famines, see the text of lecture on the subject delivered before the Royal Institution" For as informative account of 'Great Britain' on 18th May 1877 by Lieutenant General R. Strachey and reproduced with slight modifications in Appendix 1, Famine Commission (1880) Report, pp. 1-7.

8. In certain localities in this belt, it far exceeds even the latter limit; thus in Cherrapunji in the Khasi hills, the annual average is 485 inches, in Matheran, an isolated hill near Bombay, 256 inches, and in Mahabaleshwar on the Western Ghat, 252 inches.

9. After the introduction of the Constitution of Free India and the Reorganization of the States constituting the Indian Union, names and boundaries of the former British provinces have undergone many changes. Throughout this study, however, old administrative names of different provinces and areas have been retained for purposes of uniformity and for convenience of direct reference to records and official documents.

10. *Statistical Atlas of India* (1895), p. 21.

11. This is particularly true of the rice regions in eastern India (excluding of course, Assam and parts of East Bengal where rainfall is so abundant that it is hardly known to have failed

ever). While the average annual rainfall in these parts is much more than what it is in the precarious tracts of the country, the dependence of the people on a single crop for their food supply makes them peculiarly liable to famine.

12. Ganguli, B.N., *Trends of Agricultural Production and Population in the Ganges Valley*, pp. 288-289.

13. A local saying nicely sums up (in English translation) the situation as follows: The half of Adra passes dry. No rain in Hathia brings relief, If rainless Magh and Magha fly. The husbandman is plunged in grief.

    (Note: Adra is from 21 June to 3 July, Hathia 21 September to October, and Magh 31 December to 29 January). Rain is also essential in July for helping the seeds to sprout. Another saying on his subject in the same tract runs as follows: Adra barse, punarbas jae, Deen ann kou na khai. (If there is rain in Adra, but none in punarbas- 4th to 18th July – the poor will not get food to cat.)

14. *Famine Commission Report* (1880), Vol. III, Appendix 1, "Conditions of the Country and the People," p. 33.

15. Mann, H., *Rainfall and Famine (A Study of Rainfall in the Bombay Deccan, 1865-1938)*, pp. 33-34.

16. Earliest references to famines in India are to be found in the Rigveda. In the third book of the Rigveda in hymn 8, a prayer is offered "to drive poverty and famine far from us" while in hymn 53, of the same book, the danger of famine having been dispelled is mentioned. A similar prayer is offered in hymn 55 book VIII where Indra is invoked to keep people free from famine. Among other books of the earlier times which contain reference to famine, mention may be made of Nirukta by Yaska (Chapter 11, Sec. 11) belonging to the late Vedic period, *Ramayana* (Ualmik, Chapter 9, Verses 9 and 10), *Jataka Stories of Buddhists* (Jatak XXII) and Chanakya's *Arthashastra*.

17. Loveday, *History and Economics of Indian Famines*, p. 25 (also Appendix A).

18. Digby, *Prosperous British India*, p. 123.
19. *Report of the Famine Commission* (1991), p. 1.
20. Loveday, *op. cit.*, is critical of the views of William Digby and R.C. Dutt, who blamed the British Rule for the increase in the frequency of famines. But he account of the previous famines that he gives in Chapter 1 of his book only proves that famines in earlier times caused severe distress and great sufferings to the people. He has, however, nothing to say about the increase in the frequency of famines during the period of British Rule which forms the main basis of the critics attack on the British Administration of the country. The title of Chapter IV of Digby's Prosperous British India, which deals with the subject, runs as follows: "Famines: Their Present Frequency and the Cause of that Frequency."
21. Report, Part 1, para79.
22. Quoted in *Famine Commission (1898) Report*, Chapter II, para 52.
23. *Ibid.*, para 585.
24. *Ibid.*, para 585.
25. The Government records of the time on the subject contain frequent references to hoarding and cornering of supplies of food grains by the traders and speculators. Among the published reports, a good account of the feverish speculative activity In grain in times of famines is to be found in Baird Smith's Report on *Effects of Famine on Manchester Import Trade in North Western Provinces* (1861) and Vaughan Nash, *Great Famine* (1901).
26. For the effects of plenty or scarcity, in the price of corn, are incalculably greater than in proportion to the increase or decrease of quantity. Vide Ricardo, *An Essay on the Influence of a Law Price of Corn on profits of Stock*, Vol. IV, pp. 28-29.
27. Colvin's Memorandum being Annexure to Letter No. 217 dated 14 November 1860, from Magistrate Muzaffarnagar to Commissioner Meerut Division.

28. *Ibid.*, p. 50.

29. *Report on Effect of Famine on Manchester Trade,* dated 8th May 1861, para 31.

30. *Report of the Commission of Famine in Bengal, Bihar and Orissa*, 1867, part III, para 63.

31. Vide Henvey, *Narrative of Famine in NWP*, 1871. The Census was taken on 2 September 1869. Proceedings Home (Public) No. 102 (Famines), December 1871.

32. *Ibid.*, p. 170.

33. *Famine Proceedings*, No. 73, June 1900. The Census was taken in January 1900.

34. One is tempted to argue on the basis of these figures that while in the sixties the "cultivators" class was generally able to look after itself during the famine either from their own resources or with the help of credit from the money lender, by the end of the century their economic condition has deteriorated so far that they constituted the largest number among the victims of the famine. But this conclusion must be qualified by the statement that first the data available on the subject is too inadequate to establish a firm basis for deduction and secondly that statistics relating to the two periods, 1869 and 1900, are not strictly comparable as they relate to different regions of the country. There is yet another point to be kept in mind about the statistics of 1900. Bombay Presidency, to which these figures relate, had suffered from a terrible famine in 1896-97 and that of 1899-1900 followed so closely on the heels of the first that cultivators did not have much time to recuperate before they were faced with another calamity.

35. The critics of the Government generally confused ordinary poverty of the masses with the particularly straitened circumstances of these classes. Thus while the "drain" theory and the high assessments of land revenue explain the general poverty of the masses, they cannot, by themselves, provide an adequate explanation of the economic difficulties of the particular sections of the population which suffered the most from famine.

36. For a detailed account of the social and economic structure in the country at the time of the British conquest see George Campbell, *Modern India,* Chapter III, "The Institutions and Situation to which we Succeeded."

37. *Ibid.*, pp. 87, 94.

38. *Ibid.*, p. 98.

39. The village communities in Bengal at the time were the weakest in India and the institution of zamindari was widely prevalent in the province. The zamindars were of recent origin. In the North Western Provinces there were many perfect democratic communities but some village zamindaries were to be found even here. In Gujarat, the institution of zamindar was also in existence. *Ibid.*, p. 95.

40. As the famine Commission of 1880 remarked: "It be expected in every forward movement in the education of a people that while the result is beneficial to the country as a whole, some classes or individuals will fail to display the qualities needed to benefit by the advantages offered, and will fail to display the qualities needed to benefit by the advantages offered, and will suffer inconvenience under the novel circumstances to which they are unable to adapt themselves. *Report*, pp. 130-131.

41. *Economic Development of India*, p. 5.

42. Balkrishan, *op. cit.*, Chapter X, for a detailed account of the measure of supremacy of Indian manufactures in the British market and legislation passed in England against import of Indian textiles in that country in the first quarter of the eighteenth century, see P.J. Thomas, *Mercantilism and East India Trade,* London, 1926.

43. Interesting evidence of the remarkable progress of imports of Manchester goods over the period of 25 years, 1837-38 to 1860-61, is available in Baird Smith's report on the *Effects of Famine in the Famine Affected Tract in NWP* in 1861. According to Baird Smith, in Azam Garh district in 1837-38 the number of those using foreign cloth was negligible while in 1860-61, 77

percent of the population in the chef subdivisions of the district were clothed in English cloth and even "in the stronghold of the native weavers where upward of 4,000 looms were at work in 1836, 62 percent of the people now wear English clothes." In another district, namely, 1858 to 1861, or an increase of almost 100 percent over the average of the first four years of the decade. Two classes of village artisans, namely, carpenters and blacksmiths, seem to have suffered the last. Not only were they able to retain their custom as their.

44. Vide R.P. Dutt, *India Today*, p. 224.

45. Vide *Modern India*, p. 65.

46. S. Patel, *Agricultural Labor in India and Pakistan*, pp. 18-19. Patel points out that figures an overestimate "owing to the inclusion of some non agricultural labourers" This indeed seems to be the case, for the figure for 1881 instead of being higher was actually lower at 15 percent. In 1891 it was still lower at 13 percent but this may probably be due to the fact that it represented the proportion of tot and not merely that of the adult male workers as was the case in 1872 and 1881. In 1901, the figure was put at 25.1 percent. If the trend was towards increase of the proportion of agricultural labor over this period, there is nothing to explain the reversal of the trend in 1881 and 1891. It is obvious, therefore, that in 1871-72 as Patel argues some categories of non-agricultural laborers must have been counted as agricultural laborers. "See Report on the Famine in Orissa, Bengal and Bihar", part III, Para 63, the cession of the district to the British.

47. "Nothing is plainer", wrote Thomas Munro, principal Collector of the Ceded Districts (NWP), to the Board of Revenue (Bengal) in 1807, "than that landed property has never existed in India except on Malabar Coast." Other evidence on the subject is to be found in the Report of the Deccan Riot commission, para 41; S.S. Thorburn, *Musalmans and Money-lenders in the Punjab Provinces* by R.W. Cox and H. Trucker dated 13th April 1808, para 67.

48. *Report*, para 46.

49. *Report on Famine*, part II, para 48.
50. *Ibid.*, para 60.
51. Published in 1886.
52. Moral and Material progress Robert (Decennial), 1901-02, p. 449.
53. S.S. Rahavaiyangar, Memorandum of the progress of Madras presidency during the last 40 years, published in 1892. Raghavaiyangar asserted about that presidency that while the indebtedness had increased, to the question whether the agricultural classes are generally more in the hands of sowkars professional moneylenders the answer must as decidedly be in the negative p. 258. In that Presidency, the moneylending class arose out of the agriculture castes and transactions in money advance, sale of land, and trade in corn main consequently confined to castes. Moral and M. Vide Progress Report Decennial, 1901-02, p. 488.
54. For the close relation between the property Laws introduced by the British and transfer of land to the non-agriculturists, see Baines' Note on the subject in moral and material progress Report (Decennial), 1901-02, pp. 434-436.
55. *The Land System of India*, Vol. LXXV (1863).
56. *The Wealth of Nations*, Book I, Chapter VIII (Modern Library Editing), pp. 64-65.
57. Rev. J.T Sunderland writing in the New England magazine for September 1900 (Vol. XXIII, No. 1, Boston, Massachusetts) on the causes of Indian famines stated: "It has often been pointed out that the British Government of India is the most expensive Government in the World. The reason is it is the Government of the Foreigners.
58. Vide R.C. Dutt, *Economic History of India*, p. 399.
59. Minute dated 26 April 1875.
60. Quoted in R.C. Dutt, *Famines in India*, Appendix O, p. 197.
61. Speech in the Imperial Legislative Council dated 4 January 1886 in the debate on Income Tax Bill.

62. The nationalist opinion in the country was in favour of taking this step, especially during periods of scarcity and famine, but the Government steadfastly refused to do that.

63. See Atkinson, 'Rupees Prices in India, 1870 to 1985', *Journal of Statistical Society*, September 1909, pp. 496-574. Atkinson while strongly opposing the view which was later taken by K.L. Dutt in his report, "Rise of Prices" (1941) admitted that prohibition of exports would have kept down the price level in food grains.

64. As early as 1856 referring to the condition in North Western Provinces Colonel R. Strachey wrote, vide Report on the Iron Works in the Kumaon, Bhabur, 1956: "I shall probably err in the right direction when I say that one-third of the population in many districts of North Western Provinces is not only wastefully employed, but, that it actually impoverishes the remainder. The crops raised by the existing one-third of the agricultural population could, with-out doubt, often be raised as well if one-third of that population were struck off."

65. *Report of Famine Commission on Bihar and Orissa*, 1867, part III, paras 25 and 63. Referring to the rise in prices of food grains during famine in India the Commission wrote (para63): "We must repeat a doubt whether the laboring classes of England or France could withstand a general enhancement of prices of food to twice of thrice its ordinary price as do the native of India."

66. S. Devereux, *Famine in the twentieth century*, IDS Working Paper 105, Brighton: Institute of Development Studies, 2000, p. 4.

67. A.K. Sen, *Poverty and Famines: An Essay on Entitlement and Deprivation*, Oxford: Clarendon Press, 1981.

68. Quoted in P. Howe, 'Reconsidering Famine', *IDS Bulletin*, 2002, vol. 33, no. 4.

69. A. deWaal, *Famine that kills, Darfur, Sudan, 1984 - 85*, Oxford: Clarendon Press, 1989.

70. J. Dreze and A.K. Sen, *Hunger and Public Action*, Oxford: Clarendon Press, 1989.

71. *Ibid.*
72. *Ibid.*
73. *Ibid.*
74. *Ibid.*
75. *Ibid.*
76. *Ibid.*

Chapter IV

Impact of Famines on Agrarian System of Rayalaseema

The famine of 1866 extended in the South to several districts of Madras Presidency, which included Gangjam, Bellary, Salem, Coimbatore, South Arcot and Madura. Except in the first named district where stress was felt early in February, the distress in the presidency arose from the rise in prices as the result of export of food grains rather than from absolute failure of crops or cessation of agricultural employment. Rainfall in these districts of Madras in 1866 was delayed. It did not start till late in July, abundant rain fell in August even in Ganjam and distress began to ease off.

The presidency exported by sea, in 1865, one million cwt. of food grains. As against this usual quantity of export on the past, Madras had every year imported food grain from the neighboring provinces principally from Orissa, through Ganjam district, for feeding its own population. The total imports in 1864-65 amounted to 1,400,000 cwts. But in 1865-66 they fell to 1,100,000 cwts. And in the next year came down still further to 8,15,000 cwts.[1] The shortage of supplies combine with the delay in the arrival of rains in the presidency created severe scarcity and raised prices to famine levels. Thus in Coimbatore rice sold in September 1866 at 4½ seers, in Salem at 5½ seers, and in parts of North Arcot and Trichonopoly at 6 seers a rupee. In Bellary, ragi sold at an average price of 12 Lbs. per rupees in September as against the average price of 41 Lbs. In the previous year, in Ganjam, where on account of the cessation of agricultural operations, as the result of failure of rains in 1865, the conditions were those of famine rather than a mere pressure of high prices, the highest recorded price for ragi was 15 ½ lbs. to a rupee in August 1866. In spite of high prices and severe scarcity in some of the districts, the exports from the presidency continued, again merchants Seeding their grains to other places where prices were higher or equally high I preference to selling at home.[2]

Except in Ganjam and Bellary districts, there was not much this location in agricultural employment in the Madras presidency what was required to relive the distress in this part of the country therefore, was a strict government control over exports and price the relief measure adopted by the Local Government, however took the usual from the employing able bodied sufferers on relief works and distributing charitable relief to those unable to work a total of 1,82,425 persons found employment on relief works from April 1866 to June 1867 in thirteen districts of the presidency. But the relief works were not organized in right earnest till August 1866, when distress became acute the total number employed in the three preceding months, April to June, being 1,358 only. The highest number on relief works in any month during the famine was in January 1867 when 33,019 persons got this form of relief. The wage rates were more liberal than in Bengal and Bihar. In Bellary, wages were at first 3,2 and 1 ½ annas respectively for man, woman and child. In Salem and North Arcot, 2 and 2 ½ annas were respectively paid to the famine labour. Simultaneously, gratuitous relief was organized for the relief of those unable to work. A total of 5,02,279 persons were given this form of relief over a period of 15 months from February 1866 and May 1867, or an average of 33,485 per month. The total cost on gratuitous relief was Rs. 5,08,068, while the cost of relief on public works was Rs. 6,60,562 or a total of Rs. 11,68,630 of which the cost to the Government was about Rs. 10 lakhs the rest having been raised by public subscription. The total mortality from the famine in the presidency was estimated at 4,50,000.[3]

Impressed as he was by the novelty of the situation when an area suffered from high food prices and famine conditions ever when there was no widespread drought and crop were more or less normal,

Mr. Dalyell, who was appointed to enquire into the circumstances of this famine, in his Report raised important issues with regard to import of food on Government account, control of food prices and opening of supply depots for sale of food grains to the people at cheap rates. He quoted from Adam Smith and J.S. Mill to prove that these writers had not advocated, as was generally believed a policy of non-inference with private trade in grain in a situation like the one that existed in Madras in 1866-67. On the contrary, he pointed out, the most effective means that could be adopted by the Government of the relief of distress when prices were rising beyond the limits of those prevailing in the neighborhood, and the grain merchants were not importing adequate quantities into the famine area, was for the Government to create account or by offering a bounty on imports, or making advances to merchants, or any similar measure which would have the effect of throwing a supply of grain in to the distressed county.[4]

Famine in Madras

The troubles in Madras began as far back as 1875 when due to the failure of the southwest monsoon in parts of the Presidency, particularly in the eastern district of Bellary, famine conditions prevailed on a small scale. The southwest monsoon of 1876 again proved a complete failure and by August 1876 it became apparent to the local government that the impending calamity was of a great magnitude. An area of 83,000 square miles, containing a population of 19.4 million, was affected the distress reached the maximum intensity in the curliest months of 1877, and the number of relief workers touched the highest point a total of 5,35,000 tons of grain was imported into the presidency from the beginning of the trouble till may 1877, prices began to fall little as summer months advanced, but the southwest monsoon though not so deficient in 1877 was yet fitful and irregular. In August, things looked bleak and there was a fear that the north-east monsoon might also prove deficient. The viceroy visited the presidency in that month with a view to reorganizing relief works and consulting with the local government on relief policy, but with the arrivals of rains immediately after, people began to retain to their lands. The distress, however, continued in some parts of the presidency till the early months of 1878 due to low stocks and high of food.[5]

The system and policy of relief adopted by the madras government gave rise to number of controversies. In the first instance the provincial government fearing shortage of food in the presidency and the rise of prices due to withholding of stocks by the traders, arranged with messrs Arbuthnot & Co. to hold 30,000 tons of grain for the government. The reason given by the government for this action made interesting reading, the rise (of prices) was so extraordinary, and the available supply, as compared with well-known requirements, so scanty that merchants and dealers, hopeful of enormous future gains, appeared determined to hold their stocks for some indefinite time and not to part with the article which was becoming of such unwonted value. It was apparent to the government that facilities for moving grain by rail were rapidly raising prices everywhere, and that the activity of apparent importation and railway transit, did not indicate any addition to food stocks of the presidency ....retail trade up-country was almost at a standstill. Either prices were asked which were beyond the meant of the multitude to pay, or shops remained entirely closed.[6] The strategy of the government was to keep a reserve and throw it into the market wherever traders indulged in hoarding or prices show a tendency to rise beyond the levels reached in neighboring markets. This policy of keeping reserves of food for such emergencies, however, met the disapproval of the government of India on the ground that is constituted an interference with private traders.[7]

The size and nature of relief works also gave rise to the different of opinion between the central and the provincial and the government. The provincial government advocated the construction of large work of permanent utility but the Central government wanted small retail Works to be opened to begin with. The Provincial government's views were ultimately allowed to prevail in the matter.

More important than this, was, however, the divergence of views on the policy regarding relief wages. When Sir Richard Temple first visited the presidency on 14th January 1877, he found one million labours on the relief camps in the three districts of Bellary, Cuddapah, and Kurnool which had a total population of 4 millions this excusive portion of 25 percent of the population on relief works, was attributed to what he called an important circumstances which then became noticeable, namely, that "rural classes who are not yet pressed by want, who are able to sustain themselves, and to live without government aid, will largely resort to relief work so long as those works are close at hand, as the labour is easy and the wage worthwhile." In the rural classes, besides others, he included "pattadar riots that is the peasant proprietor, riots having beneficiary interest in the soil under ryotwari settlement, and paying land revenue to the government." A family not in need of relief would be able to earn Rs. 10 per month (Rs 3¾ for man plus 2¾ for woman and 3½ for two elder children) or Rs. 40 for 4 months, which is a considerable amount for a rustic family and would help it in digging or deepening an irrigational well, replacing dead cattle and so on. These were, according to him, undesirable results of Madras Relief Policy which made "public charity indiscriminate." He suggested, therefore, that fresh admissions to relief works should be stopped, the existing labour force on relief camps should be properly screened with a view to eliminating those who are not in need of relief, to reduce the wage from 1½ lb of grain to 1 lb of grain (from 2 annas to 1½ anna per day) and to pay wages everywhere in cash instead of grain. The 1 lb ration on relief camps, nick-named "Temple Ration." Had the effect of driving away large numbers from the relief camps. The number of relief works fell from 9,07,316 in the first week of February to 6,62,195 in the last week of march but the numbers "too weak for work requiring cooked food in the relief houses" increased from 38,163 in the first week of march to 99,113 in the last week of the same month.[8] The complaints about the inadequacy of food in relief camps resulting in heavy incidence of diseases due to malnutrition and exhaustion and increased mortality reached London and the secretary of State advised the Government of India not to place too much restriction on the discretion of the local government.[9]

Under this pressure from Home and the growing criticism of the inadequacy of wages especially from medical experts, the government of India ultimately relented and permitted the local government in May 1877 to revise upward wages of the famine labour.

The number on relief camps was over one million in January 1877. The introduction of stricter discipline, higher task and the lower wage scale in that month had reduced the number for the time being, but shortly afterwards, it began to increase again (especially of people on gratuitous relief), and was over a million in May 1877. In September 1877, it reached the maximum figure of 2,218,000. Of this over 1 million were labourers on works, the rest receiving gratuitous relief. The total expenditure on relief works was Rs. 94,94,369 in 1876-77 and Rs. 230,54,463 in 1877-78 making a grand total of Rs. 325,48,832 in the two years.[10] Simultaneously Rs. 170,33,889 were spent during the same period on gratuitous relief. There was terrible loss of life in the Presidency as a result of the famine. The Famine Commission estimated, on the basis of comparison of census returns of 1871 and the test census taken in 1878 that the total loss was 31/2 million lives.[11]

An interesting feature of the Relief Policy of the Government was the liberal money advances made to the distressed population for the purchase of seeds and for the construction of wells and tanks. The advances during the famine amounted to over 18 lakhs of rupees. Another indirect form of relief was revenue remission. While the Madras government was in favour of remitting a large portion of land revenue, it was compelled under pressure from Sir Richard Temple to limit remission to the minimum and grant instead suspensions only. The total loss of revenue to the government as the result of the famine was Rs. 150 lakhs.

Famine in Mysore

The native state of Mysore was under British regency at the time when the failure of the southwest monsoon in 1875 resulted in heavy damage to the autumn crop and later also to the winter crop. There were signs of distress among the people of the State but it was not so acute as to receive the attention of the Administration. Since the agricultural security was provided by the Tanks, wells and irrigation channels in Mysore, this state had an advantage over the neighboring British Province of Madras and it was able to tide over the difficulty for the being human. Due to the failure of rainfall for the second year in succession in 1876, the tanks and wells dried up to and there was severe distress and high mortality both among man and cattle. The total deaths from famine and cholera in two years were estimated at 1.05 million[12] out of a total population of 5.05 million.

There was a sharp rise in prices of food grains, the price of ragi, which was a staple food of the people of Mysore, rose to 22 seers per rupee, or more than double the usual price, in November 1875, to 18 seers in August 1876, and to 6½ seers in December 1876,[13] from November 1876 to July 1878, a total of 2,37,179 tons of food grains were imported from the state by private trade. No attempt was made by the Government to import food grains on its own account or control prices.

The relief measures were belated. The Commission of Regency was found to be absolutely unprepared when the famine came and they displayed an utter lack of sense of urgency in the initial stage. As one writer put it at the time, "there was the usual bewilderment, the usual vacillations and it was not until the evil had reached the climax that it was seriously grappled with."[14] When the Government understood finally that was required, there was no holding back, but by that time it was too late. A total of 1,069,290 were found work over a period of 20 months, December 1876 to July 1878, or a monthly average of 53,470. In October and November 1877, the average was 75,000. Simultaneously an average of 90,575 persons were afforded gratuitous relief for 11 months from December 1876 to October 1877 and an average of 1,23,000 for the next 9 months at the total cost of Rs. 21.25 lakhs.

Famine in Hyderabad

The drought of 1876 extended to the southern half of Hyderabad territory that is to the tract lying in the angle between the famine districts of Bombay, Ahmednagar, and Sholapur, Belgaum and Dharwar on one side and the famine districts of Madras, Bellary and Kurnool on the other. A total area of 30,708 square miles containing a population of 1.9 million was affected, of which 6,942 square miles with a population of 7,29,000 were affected severely. There was, however, a rich harvest in the northern part of the sates which continued to export grain both to Bombay and Madras. In 5 months ending with January 1877, 40,000 tons of food grains were exported from the state.[15] Prices raised under the influence of local scarcity of *jowari* rose in January 1877-79 seers to a rupee. Conditions deteriorated as the summer advanced, and July and August 1877 were the worst months. The outlook improved with the arrival of rains in September.

With the onset of the famine, instructions were issued to the effect that "no interference was to be permitted with prices of food grains"; "useful public works were to be opened in districts where relief was wanted" and ordinary works in progress at the time, in other parts, were to be stopped. Weekly reports were called for from all districts on points indicative of the conditions of the country.

Relief was provided, as in the British territory, by offering employment on relief works and feeding destitute in poor houses. On an average, 18,290 persons were employed on relief works for 11 months. The total cost on this form of relief was Rs. 8,38,122, a daily average of 6,471 persons received free food in the poor houses for 10 months at a total cost to the State of Rs. 2,44,347.

Scarcity of Rainfall in Madras Presidency 1884-85

In Madura, Coimbatore and Cuddapah, the rainfall for the period from 1st April to 30th September was, in 1884, nearly one-half less than the normal, in Bellary and Tinnevelly more than one-half less, and in Anantapur nearly two-thirds less. Other parts of the presidency were not affected but, in the affected districts, the failure of southwest monsoon was more complete than in the famine of 1876. Writing in the middle of October the madras Government described the position as "vritical."[16] There were, however, copious rains in November and December in all districts and the Government's anxiety disappeared excepting in respect of Anantapur and Bellary where "the comparatively high prices pressed heavily on people accustomed, as are the people of Bellary and Anantapur, to cheap food."[17] No attempt was made by the Government to control prices or divert supplies of food grains to the famine districts but it attempted to relieve distress by providing work to those who required it. The Government was not ready, when the famine came, with any scheme of useful work, and employment had consequently to be provided on earthwork which was continued till June1885 when the monsoon set in. The Government described the situation in the two districts as "a state of distress not amounting to famine."[18]

Scarcity in Central Provinces, 1886-87

The rice crop in Bilaspur and Raipur districts of the Central Provinces was damaged in 1886 on account of a long break of rains August and September. During the year the rainfall in the Central Provinces averaged 29 inches as against the normal of 40 inches.[19] The deficiency extending to Mandla, Seoni, Narsingpur, Hoshangabad, Betal, Chindwara, Wardha, Balaghat and Bhandara districts. The Chhattisgarh division was the worst affected. Distress was feared in the two districts of Bilaspur and Raipur in the hot months of 1887 but it did not appear. The explanation given was that the tract had great resources, that the previous season was good and good and consequently there were ample stocks in the villages, and that though prices were higher than usual, rice selling at 20 to 24 seers against 30 to 40 seers in 1885, they reached the scarcity level. Relief measures of the Government were confined to the offer of work by District Boards on digging of tanks.

The state collected its full revenue demand, the total collection in the year of scarcity amounting to Rs. 62.57 lakhs as against Rs. 60.60 lakhs collected in the previous year.

Scarcity of Rainfall in Madras, Bombay and Bengal, 1890-92

With the exception of Central Provinces, North-Western Province and Oudh, all other parts of the country had deficient rainfall in 1891. The overall deficiency for the country as a whole was 9 percent but is the Madras Presidency it was 15 percent, in Bengal and Bihar 13 percent, Bombay 14 percent, and Hyderabad 25 percent. In Madras, rainfall was deficient both in 1890 and 1891 in Bengal deficiency was experienced in 1891 and 1892 in Bombay 1891 was a bad year but the monsoon both in 1890 and 1892, more than the average rainfall. Local scarcities and famines were, therefore, experienced in different parts of the country over the three years 1890 to 1892.

Famines in Madras, 1890-92

The Madras Government had hardly wound up its relief operations in Ganjam when it found itself faced with deficiency of rains throughout the presidency in 1890 and1891. The principal areas affected by drought were Cuddapah, Nellore, Chingleput, North Arcot and Coimbatore which suffered the most in the first six months of 1891. At the end of February 1892, the affected area in the presidency amounted to 31,069 sq. miles covering a population of 4.98 millions.[20]

As a result of partial failure of crops there was a sharp rise in prices of food grains. Rice in 1891-92 was only 10.5 seers to a rupee, as against the average of 15.6 seers in 1887-88, Ragi was 17.8 seers and 30.4 seers respectively and cholum 16.88 seers and 27.9 seers. No action was taken by the Government to check the rise in prices but the Ganjam famine had taught a lesson that the failure of crop of a "single season was sufficient to cause extensive distress a fact which had formerly been doubted."[21] Early steps were, therefore, taken at organizing relief and a total sum of Rs. 16½ lakhs was spent by the Government for the purpose. "Charitable" relief was administered to those unable to work in the form of cooked food which people were very reluctant to accept.[22]

The famines were accompanied by cholera which took its usual toll among those suffering from deficiency of food. The death rate in the affected districts was higher than the normal in Kurnool 44.1 per mile against a normal of 31.8 per mile, in Bellary 32.2 against 28.7 and in Anantapur 29.4 against 26.2.[23] But the famine Commission did not consider the increased mortality as "abnormally high", from which it deduced that "famine was successfully managed" at not an excessive cost. But the management of a small local scarcity which causes a mortality of about 45,000 can hardly be described as "a success."

Famines of 1896-97

The famine of 1896-97 was more widespread and severe than any the country had experienced before. It was officially described at the time as the most disastrous famine of the century. [24] Besides the four Bundelkhand districts of Allahabad Division where famine conditions had already prevailed since 1895, the famine of 1896-97 affected, more or less, every part of India. Among the provinces affected were Bihar, Bengal, Bombay Deccan, the Deccan districts of the Madras Presidency, the North Western Provinces and Oudh,[25] the Central Provinces, the Punjab and Burma. Distress was particularly severe in the Central Provinces, the North Western Provinces, Hissar district in the Punjab, and the Deccan districts of Bombay presidency, Narrative [26] of this famine put the total area of the famine track at 504,940 sq. miles and the population affected at 96,931,000 persons. If, however, we exclude native states and take only British India including Burma, the famine tract covered an area of 224,828 sq. miles and affected a population of 62.4 million.[27]

The immediate cause of the famine was the failure of the autumn rains over a large part of the country. Over a wide area there was no rainfall from the middle of August of the middle of November a period during which India gets normally about one-third of her annual rainfall. The autumn food crops such as millets, pulses, and rice was badly damaged. In the Central Provinces the crop yield for the year was estimated at 45 percent of the normal, in NWP and Oudh 60 percent, in Bengal 67 percent, in Bombay 65 percent, in the Punjab 75 percent and in Madras 80 percent.[28] The loss of crop in the year was estimated ot one-third of the Average annual production of "18 or 19 million tons." Against this the total imports from Burma during the period of famine amounted to 6 lakh tons. The deficit had therefore to be meeting almost holy from accumulated reserves, if any, or by a reduction of consumption under the stress of high prices. In the North Western and the Central Provinces, the famine had been preceded by short crops in the previous one or two years and in the former by a famine of considerable severity in the Allahabad Division. Food stocks in these provinces were naturally low at the time when the famine came. In Bombay,[29] Bengal[30] and Madras,[31] on the other hand, heavy exports of grains in the previous years had drained away whatever surplus there might have been. Exports continued unabated from the Bombay presidency even up to September 1896 when the famine was found the corner, with the large shortfall in the out-turn of food grains and in the absence of any large stocks accumulated from previous years, the country was faced with a food shortage of unprecedented magnitude.

The first intimation of famine came to Government from north Western province at the end of September 1896. Famine took a violent turn next month. There grain riots in Delhi, Agra, Nagpur Muzaffarnagar and Mhow. Grain shops were looted in Mhow on the morning of 15 October 1896. There was civil commotion and unrest in Bombay against continuing exports of food grains from the presidency at a time when people faced the threat of famine. There was a growing demand for imposing a ban on food exports and for encouraging imports of food from aboard on Government account. The Government of India, however, refused to change it food policy[32] and steadfastly clung to the view so far held that, "even in the worst conceivable emergency so long as trade is free to follow its normal course, we should do far more harm than good by attempting to interface."[33] the policy of non-intervention in trade was carried to absurd limits. Thus the Bengal Government was censured for a proposal made by it of making advances to landlord and traders to import food grains,[34] while in Bombay some municipal committees which tried to open fair price shops were reminded that, under the Local Authority Funds Act, the committees were not authorized to use their funds for meeting " any loss incurred or in interest due on sums borrowed for the supply of grains."[35] The only exception to this general policy was the importation of grain on Government account in such remote area as Palamau district in Bengal, the Bhadrachalam Taluk of the Godavari district in Madras, the Shetpal relief work in Poona district in Bombay, and in Mandla and Balaghat districts in the Central Provinces.

Barring these minor cases, in the country as a whole the adjustment of the reduced supply to the demand for food grains was left to the operation of the price mechanism. It was natural, under the circumstance, for prices of food grains to shoot up. According to the Famine Commission of 1880, "in times of very great scarcity prices of food grains rise to three times their ordinary amount, so that were as in ordinary years the price of food grains of the mass of the people may be from 20 to 30 seers a rupee, in times of great scarcity, it will rise to 8 to 10 seers a rupee, or even higher." The course of prices during the famine of 1896-97 was in accordance with the past experiment.[36] For example, in Darbhanga the common rice was 24.21 seers a rupee in 1895, 13.25 seers in October 1896 and 9 seers in June 1897. In Bellary district (Madras) Cholum was 29.7 seers a rupee and ragi 30.35 seers a rupee in 1894, but by June 1897 the former was selling at 10.7 seers and the latter at 15 seers. In Cawnpore (NWP) wheat was 17.9 seers a rupee in 1894, and 8.12 seers in November 1896.

Extension of Irrigation

Within the limits of the finance available, the choice of irrigation schemes for execution by the State had so far been guided by two main considerations: (i) the development of the country's resources, and (ii) the financial productiveness of the investments in irrigation works. Protective works, that were important for their value as providing insurance against famine, could not be financed out of the current revenues of borrowed funds. The Famine Insurance Fund, which was the only source of finance for such works, was too me agree to support any considerable programme of protective irrigation works. The irrigation policy of the Government thus bore no relation, till the end of the nineteenth century, to the requirements of the situation. The appointment of an Irrigation Commission in 1901 was, therefore, not merely an indication of the readiness of the Government to devote greater attention, in its future programmes of public works, to schemes of irrigation it also marked a fundamental change in the thinking of the Government on the purpose of irrigation works in India. This was clear from the instructions of the Government. "The main question as regards new works", the Commission was told, "is not whether they will be likely to prove directly remunerative but whether the net financial burden which they may impose on the State, in the form of charges for interest and maintenance, will be too high a price for the protection against famine which they may be relied on to afford, and it is from this point of view that we should consider proposals for the extension of irrigation in districts in which it (rainfall) is very insecure or precarious." this of course, did not mean stopping further extinction of productive works which the Government constructed "as a matter of business."[37] It only marked the beginning of a move in the direction of providing irrigation facilities to those large areas in the country, including some of the portions which were most liable to famine, "where no works of productive class are possible, and which must be protected, if at all, at a certain cost out the revenues."[38]

The Commission found that it was something of a physical and financial impossibility to protect all parts of the county against failure of crops and famine. In areas like the Deccan districts of Bombay and Madras, the Central Provinces and Bundelkhand, where protective irrigation was most urgently required, there was "not a prospect of any new irrigation works on any considerable scale proving directly remunerative." In some of these construct works of irrigation, the cost would have been prohibitive. There was, the Commission felt, "a limit to the expenditure which may be incurred and a risk of imposing a burden on the country which may be even greater than the famine itself."[39]

However, the Commission pointed out that irrigation works had not in the past received the attention that they deserved, that there was still great scope, in some parts of the country, for the extension of irrigation works which would directly prove remunerative, that there had been over the last 25 years very little increase in the area covered by the minor irrigation works which had suffered from comparative neglect, and that, besides extension of State protective works, much could be achieved by encouraging construction of private irrigation works, such as well, through State advances and revenue concessions. A twenty-year irrigation plan, involving a total expenditure of Rs. 44 crores, was drawn up by the Commission of this sum, Rs. 15 crores was proposed for investment on productive works, Rs. 20 crores on protective works and Rs. 9 crores on the intermediate works. The Commission also recommended that while "the programme of future expenditure on irrigation should provide for the construction of as many productive works as can be proposed, in whatever parts of the country they are situated and without reference to the urgency of protection to the locality," the sanction of a proposal for a protective works should not the "withheld simply on the ground that investments elsewhere are more profitable." Each individual scheme, it was recommended, should be considered on its merit, the criterion being that "productiveness" should receive greater consideration than "protectiveness."

The solution to the irrigation problem of the tracts most liable to failure of crops, the commission felt, lay in the encouragement of the construction of well and minor irrigation tanks which were best left to the individual or private care. But the state could stimulate the extension of these works in the country by exempting from land tax the benefits accruing to the landholder from permanent improvements in land and by granting him financial assistance to enable him to undertake these works. The Commission pointed out that in Madras "permanent exemption of improvements from taxation has justified itself as an effective encouragement to the construction of well and that its trail for that purpose in other provinces, where exemption is at present only temporary, would be justified in tracts exposed to famine in which special encouragement are require." A similar recommendation had been made earlier by the Famine Commission of 1880 and a provision to that affect had been included in the land improvements, loans bills, but in the course of the debate on it in the legislative Council, the provision was dropped. The famine commission of 1901 repeated the recommendation.

The Government of India, in its comments[40] on the recommendations of the irrigation commission, welcomed the declaration made by the commission against the prevailing popular view that the "Whole of India can be securely protected against famine by the construction of irrigation works." This declaration by an authoritative body of experts, the Government felt, "Supplies an answer to those who hold that because, taking one year with another, the rainfall in India was sufficient for its irrigational needs, it is both possible and our duty to store the surplus of good year and to render failure of crops impossible."[41] The reference was obviously to General Sir Henry Cotton[42] who had died a few years earlier. The Government pointed out that so far as the extension of productive works was concerned no question of principles was involved. Their construction would be pushed forward so far as finance was available, and the Government undertook to find Rs. 220 lakhs a year of Rs. 44 crores over the next 20 years, as recommended by the commission, for irrigation works. The recommendation for abolishing the official distinction between productive and protective works, made by the Commission, was turned down, but the Government undertook to devote in future greater attention to the construction of protective irrigation works, "The real Justification for laying a burden upon the whole of India in order to protect the inhabitants of an insecure area from famine is," the Government of India wrote, "not that expenditure or pecuniary loss will be thereby avoided, but that the construction of the works will prevent an infinity of human suffering and save a number of human lives." In deciding whether to protect a certain tract at a certain cost, the Government went on to say that "the element of sentiment" could not be excluded. Quoting the example of Bombay Deccan in which protective works were not expensive and were likely to yield small net return in normal years on expenditure incurred on irrigation the Government pointed out that this part of the country was so exposed to the ravages of famine, and the distress, when it comes, is so general and so acute, that the Government is under a moral obligation to do something to protect its inhabitants if it is in any way possible.

Another recommendation of the Commission related to the question of classification of minor irrigation works. The practice of the Government of far had been to treat all such works as protective, irrespective of the amount of revenue yielded by the, and to charges all capital expenditure incurred on these works to the current revenue. Under that system of accounting, the importance of such works was liable to the minimized because some of the existing works of this class were really productive but were not being shown as such. The Commission therefore recommended. That "all existing minor works, for which both capital and revenue accounts are now kept, and the revenue account of which can be accepted as indicating, with reasonable accuracy, the true return due to the capital expenditure should be transferred to the class of major works, whether they fulfill the conditions of a productive work or not." The Commission thought that of this recommendation were accepted, "the productive and protective value" of such irrigation works would be greatly increased. The Government however, turned down this recommendation of the Commission on the ground that it was based on "the understanding that the distinction between productive and protective works would be removed." As the Government had already indicated its intention to "retain the distraction," it followed that only those minor works which had proved "productive" could be transferred to "Major Works" so as to be improved and developed from loan funds. On the general question of paying, in future, greater attention to the development of minor works, the Government admitted that during "the period of financial stress which accompanied the fall in the sterling value of the rupee, it was impossible to devote any considerable additional sums from general revenue to expenditure under this head It was, however, realized how that works of this class were of great importance to the country and the Government expressed itself in full agreement with the Commission "as to the advisability of increasing expenditure on them". As for the private irrigation works such as wells, the Government expressed its willingness to give financial assistance for their construction but in the matter of permanent exemption of land improvements from taxation, it stuck to its earlier decision embodied in the Land Revenue policy Resolution of 1902.

The Commission's recommendations and the Government's action there one could not have, in the nature of things, produced immediate results; but they paved the way to a broader and bolder approach to the problem of irrigation in India, with a view to ultimately saving the country from the menace of recurring famines.

Improvement in Economic Condition of Agriculturists

There is growing unanimity that Indian agriculture has been passing through a phase of serious crisis. The crisis in agriculture has many manifestations. If it was not possible by means of irrigation works to secure the country completely against occasional drought, it was necessary, if occurrence of famine was to be prevented, to improve the economic condition of that section of the agricultural population which was the main victim of famines, in order to strengthen its power of resistance against famine. Three principal factors which weekend the staying power and mostly accounted for the poverty of the poor peasants were their indebtedness and consequently large yearly payments of interest to the moneylender, the exaction land revenue demand of the State, and the high rental charges of the landlord. The distributive system in Indian Agriculture worked to the disadvantage of the actual tiller of the soil and the most important reform needed for improving his economic condition was regulation of the share of the moneylender, the State and the landlord in the produce of the land. Reference to indebtedness and its evil consequences was made by the successive famine Commission, The attack on the land revenue system came mostly from the critics of the Government, and the exactions of the landlord found prominent mention in the Government's Land Revenue Resolution of 1902. The period that we are surveying was marked by notable action on the part of the Government to tackle the problem of indebt tenses and limitation of the land revenue demand of the Government.[43]

The state which continued to export grain both to Bombay and Madras. In 5 months ending with January 1877, 40,000 tons of food grains were exported from the state. The conditions deteriorated as the summer advanced, and July and August 1877 were the worst months. The outlook improved with the arrival of rains in September.

With the onset of the famine, instructions were issued to the effect that" no interference was to be permitted with prices of food grains";"useful public works were to be opened in districts where relief was wanted" and ordinary works in progress at the time, in other parts, were to be stopped. Weekly reports were called for from all districts on points indicative of the conditions of the country.

Relief was provided, as in the British territory, by offering employment on relief works and feeding destitute in poor houses. On an average, 18,290 persons were employed on relief works for 11 months. The total cost on this form of relief was Rs. 8,38,122. A daily average of 6,471 persons received free food in the poor houses for 10 months at a total cost to the State of Rs. 2,44,347.

Accordingly to Andhra Pradesh Directorate of Economics reports 2014, more effects shows not only sequence of fall down of the rainfall but also drought. The observation of the report in Kharif season drought revels clearly. Almost all in Andhra Pradesh out of 664 mandals there is an abundant rainfall accorded in only in 9 mandals, and remaining 99 mandals very less rain fall was recorded. Compare to previous time in 461 mandals less rains are fall down, form the begging of the Kharif season to till date Andhra Pradesh Directorate of Economics reports are observed, the table clearly shows percentage and more or less rain fall conditions in the state:

Table 1: Detailed information of the District wise rainfall in Andhra Pradesh (Rain fall in millimeters)

Rain Fall				Mandal Zones			
District	Normal	Recorded	Percentage of Lack	More	Normal	Lack	More Lack
Srikakulam	959.8	861.3	-10.3	4	18	16	0
Vijayanagaram	910.6	877.2	-3.7	3	23	8	0
Vishakapatnam	964.5	793.4	-17.7	1	20	21	1
E.Godhavari	1025.2	526.6	-48.6	0	02	48	10
W.Godhavari	995.5	564.2	-43.3	0	01	40	05
Krishna	892	528.5	-40.8	0	0	47	03
Guntur	702.7	423.7	-39.7	0	06	43	08
Prakasham	66.8	340.9	-48.9	0	02	35	19
Nellore	741.7	347.1	-51.9	0	01	31	14
Chittor	679.4	370.5	-45.5	0	02	55	09
Kadapa	575.3	383.7	-50.7	0	1	34	16
Anantapuramu	466.1	258.0	-44.6	1	8	41	13
Kurnool	585.3	396.2	-32.3	0	11	42	1
Total A.P	**781.9**		-	**9**			**99**

Source: Andhra Pradesh Directorate of Economics, 2014.

The Rayalaseema Region of Andhra Pradesh consists of four Districts Kurnool, Kadapa, Anantapur and Chittoor. The entire Anantapur District, major parts of Kadapa and Kurnool, and parts of Chittoor, are reeling under severe drought conditions. The Rayalaseema Drought Relief Movement (RDRM) (a group of eighteen rural NGOs), and the Independent Commission for People's Rights and Development, (ICPRD), New Delhi organized a two day regional level workshop of farmers for Coalition Building to Combat drought (at Kadiri, Anantapur district). As part of a pioneering advocacy effort, on the issue of drought, the workshop made a detailed and systematic analysis of the factors/ policies leading to the severe crisis and social upheavals. Practical strategies to alleviate the severe economic stress and social turbulence were outlined by multi-stake holders (67 participants including 20 NGOs from 4 districts of Andhra

Pradesh participated in the effort). The major impacts of drought as highlighted by the report are given below.

Major Impact of Drought

- — Loss of crop - Debt bondage Suicides
- — Distress migration
- — Increase in child labour
- — Rampant corruption in food for work program
- — Fodder scarcity leading to selling livestock
- — Trafficking of women and children for sexual exploitation, labour
- — Pauperization of farmers

The Human Rights Forum claimed that distress migration has left many villages almost empty, with the exception of the elderly and young. Those who had not left their mandal (smallest administrative unit) were at the vagaries of the contractor. For example, a contractor took a group to Kerala, during which time, three women missed the bus. They resorted to begging in order to survive. Only one woman returned to the village, while the other two are still missing. As they were children of rural folk, they did not know traffic rules of the city and some of the children died as a result of traffic accidents. Language was also a barrier, e.g. some of the migrants were accused of theft and evicted without receiving remuneration as they could not express themselves in the local language.

Historically, in Anantapur, it is evident from the stone carvings in ancient temples, for example the Laxmi Narasimha Devalayam in Kadiri, and the reports during the British Government that there were many famines in the area. The agricultural conditions were unfavourable, owing to famines and the scarcity of food grains, many died of starvation. However, there is no indication of suicides during these periods. Why have suicides suddenly become rampant in the area?

In the year 2000, there were 48 suicides and in 2001 there were 93; and in 2002 there were 102 suicides. Of these, 55 were women and 5 were girls. Why are people committing suicide? Has the mental stability of Anantapur farmers weakened and, as a solution, resort to suicide? Agriculture is the only occupation in the area, though precarious, and often manifests loss and liabilities. The continuous loss of inputs and crops engender a quagmire of debts and the pressure from the moneylenders generates hopelessness.

Are the farmers avaricious? Do they want to make rapid profits and hence resort to mono-cropping of commercial crops, i.e. groundnuts? How and why did the farmer fall into this trap of mono-cropping of groundnuts and why are they not able to come out of it? Around 1990, the groundnut cultivation became popular in the district. The government encouraged groundnut cultivation with input subsidies and the price became favourable. The farmers continued the crop due to its ability to withstand low water inputs. Even without rainfall for two to three years, groundnuts can be cultivated. During this time, the government began to import palm oil and created competition, causing the price for groundnuts to fall. It is these policies that provided impetus to groundnut cultivation on the one hand and imports on the other hand - yet played havoc with the lives of the farmers in the area. The modern day famine is different from that of the old. Today, a good network of rail and roadways are available and the granaries are full. Yet people are dying of hunger. It is within this context that public interest litigation, positively acknowledged by the judiciary, led to the Food for Work (FFW) program. This emerged as a consequence of the directive of the judiciary. Two lakh bags of grains have been claimed in the district, most of which has been recycled in the godowns. Did not governments, political parties, social activists, NGOs and networks fail to advocate and ensure its access to the poorest? Anantapur has been declared a "desert area after crores of rupees have been spent on it when it was initially labelled drought prone". The failure of civil society organisations in the area is evident, he complained. Approximately 7 crore rupees are provided as input subsidy.

However, this has turned out to be a subsidy for suicides as many of those who died were using a harmful pesticide. The State Government's 'Vision 2020' in the agriculture sector is aimed at reducing the number of people dependent on agriculture as a livelihood to 7%. Mr Basha questioned whether these suicides are a means to achieve the goal of decreasing the number of farmers? Downsizing. He said that it was contractors that mainly benefitted from the Food For Work (FFW) program. The government was able to distribute rice freely and ask the farmers to work in their own fields.

Having analyzed the effects of the policies and their implementation, Mr Basha claimed that politics has become criminal and questioning policies or asking for transparency led to victimization. There are so many schemes operating in every village, how many news reporters or NGOs know the number of beneficiaries under the Annapurna scheme? For a single work, ten bills are often raised and sent; every village has at least ten schemes amounting to approximately two to three lakhs. Mr Basha argued that NGOs and civil society should ensure that the benefits reach the poorest and target groups by using the right to information act.

In the discussions that ensued, Mr Basha said that it is not easy to obtain water from the Krishna River due to a clash of interests. However, solutions in the interests of people from both areas are possible. Some areas have groundwater, but Anantapur does not, he agreed. It is necessary to convince and start discussions amongst people. Mr Basha further explained that the Bachawat tribunal allotments (for water allocations) ended in May 2000 and a new commission is being established some of the waters allotted to Andhra Pradesh have not been utilized, Mr. Basha pointed out, this may prove difficult in requesting future quota allotments in terms of higher requirements, he cautioned. Karnataka State has already tried for higher allotments (through Almetti), but the current authorities were not willing to discuss these issues, he concluded.

Vishweswarao, an activist journalist from Anantapur (Andhra Jyoti), examined the history of drought and its linkage to the current crisis of suicides. An historical assumption of the drought in the region is based on the theory that drought occurs when there is no rain, he said. Through it is not linked directly, yet, if there is no water, there is drought. "What could the government do? It all depends on the rain god, the government is not the cause of the problems of the farmers", claims the general perception, he lamented.

The rain in the region, particularly in Anantapur, is less than that of Jaisalmer or Israel. In a period of 124 years, there was a shortage of rain for 64 years. In September 2001, 1-43 tmc of water flowed through Anantapur but did not bring any change to people's lives!

History speaks of large granaries in the area; pearls were sold on the streets during the time of King Sri Krishna Devaraya. It is argued that during this period, there were no perennial rivers in the area. Gazettes are available in which the expenditure on drought may be found and the severity of the droughts can be estimated. Examples highlight the severe and remorseless nature of the droughts and famines in the area that have resulted in the heavy loss of human lives and livestock. It was these conditions during the British Colonial period that forced Sir Arthur Cotton to take up irrigation projects, he pointed out.

Mono Cropping-Cash Crops, the Technical Mission for Oil Seeds identifies necessary measures to fulfill the needs in this sector. When there was a shortage of oil seeds, the government encouraged groundnut cultivation through a number of inputs and marketing facilities in this area. A higher number of farmers initiated groundnut cultivation. At the same time, the New Economic Policy – withdrawal of subsidies came into force. This increased the input costs of fertilizers and pesticides.

Under the Open General License Policy, imports of palm oil were initiated. Even when record crops were grown, there were no remunerative prices. Generally, crop cultivation depends on local geographical and climatic conditions. However, remunerative prices and market demand also play a crucial role. Where there are irrigation facilities, the farmer can produce crops in relation to market demand. In a crisis situation, farmers are able to rely on paddy due to its market demand and can receive some form of remuneration. Other crops do not have the same effect.

However, paddy cultivation in the region is minor; rain-fed crops, such as groundnuts and grams, do not command adequate prices. The dilemma is acute, he highlighted. Scanty rains, loss of fertility, and lack of irrigation causes untold sufferings for the Rayalaseema ryots. Further, liberalization policies and economic reforms have pushed farmers to the verge of suicide, he argued. During the times of the historical figure and monarch, Sri Krishna Devaraya, digging tanks and building temples were the primary works undertaken. Tanks to store water for irrigation and temples were necessary in order to fight the unequal palegar system. The rulers of the period focused on the needs of the agricultural sector. However, it seems that the major progress made in the agricultural sector was in the 14th-15th centuries. There has been no further progress since this period, it could be said in a manner of speaking! The present day public works concentrate on roads and other types of construction. Emphasis is given to the ten percent commission system, required by the upper political echelons, (as their share) in terms of public works, he said.

Vishweswarao maintained that regional development is directly linked to the proportion of lands irrigated in the area. The state average for irrigated land is 36 percent, while in coastal Andhra Pradesh the irrigated area is 63 percent, a meagre 14.25 percent in Anantapur district, 13.89 percent in Kurnool district, 26.77 percent in Cuddapah district, and 36 percent in Chittoor district. However, this irrigated land cannot be compared to the irrigated lands of Krishna and Godavari districts, as the majority of irrigated lands in Rayalaseema are fed from small tanks and ponds, which are filled once in every three-four years. Approximately 2.67 lakh hectares of cultivable lands are lying idle in Rayalaseema - the highest in the state.

Furthermore, the area of uncultivable land in Rayalaseema is also the highest (6.84 lakh hectares) in the state. Thus, from 30 lakh hectares of cultivable land in the region, only 5.6 lakh hectares have irrigation facilities. 24 lakh hectares of land is devoid of irrigation facilities. Another misconception is about the direct linkage between trees and rains. Trees are a crucial part of the environment in attracting the rains. Despite several crores of rupees that have been spent on increasing forest cover, there is a significant lack of trees and deforestation in the area. The extent of monitoring reports indicate that the available area of the district would not have been sufficient, if all the saplings said to have been planted were alive in reality! CIDA, the donor agency, it is said has stopped funding this program citing the above issues particularly the poor survival rate as a primary reason. In the process, seventy crores of rupees have gone down the drain, he lamented. In a discussion relating to watershed projects, it was highlighted that if there are adequate rains, then these could be harvested into the lands. What use are watersheds when there is no rain?, he questioned. A mere nine percent of rainwater is retained in the soil. Therefore, if rainwater is to be utilized effectively, it should be stored during rains and used later.

However, Vishweswarao argued that this very concept is misplaced. He claims that through watersheds, corruption is decentralized and vested interests have converged together. In order to permanently solve the problems of Rayalaseema and, in particular, Anantapur farmers, he emphasized the need to provide irrigation sources through the Krishna River water. The government has highlighted the chronic drought situation to the Bachawat tribunal in order to obtain more water allocation from the Krishna River. However, this should be provided to the Rayalaseema farmers. In the past, people died due to famines but now it is due to the policies of the government; During the British period, the area was under the Madras Presidency and there still was a supply of grains and fodder to the area, he concluded.

Trafficking of Women and Girl Child

Describing the situation of women in the area, Bhanuja (REDS) said that the conditions caused by drought have played havoc with the lives of women in the rural sector. Women have always been an integral part of agricultural operations. Approximately 55 women from agricultural families have resorted to suicide. Even children have resorted to suicide due to the difficulties faced by their parents. She quoted a case study; Jayalakshmi, a 14 year old girl, was considered a very good student by both villagers and teachers. Considering the difficulties faced by her mother following the migration of her father to earn a livelihood, Jayalakshmi resorted to suicide.

The continuous failure of crops, debt trap, and suicides have all forced women to become sole breadwinners in many families. Drought has also destroyed supplementary livelihoods by making the maintenance of milch animals difficult. Bhanuja recalled that many sugali women (a tribe) in Lakkireddypally and Galiveedu mandals of Cuddapah district have been trafficked for sexual exploitation due to loss of their livelihood. Personal values have been eroded to such a degree that families have begun to accept this as a livelihood option, she lamented. Many trafficked women have gone to Pune, Bombay and Delhi. Locally they are involved in such activities in nearby towns, highway dhabhas (wayside cafes) and few brothel houses. As a result, women headed families have to carry their children along with them. The children are also often perverted from the child hood in terms of sexual attitudes. A majority of sex workers are affected by the STD, HIV/AIDS, leading often to high mortality rates, amongst this group, she said.

Similar situations are occurring and increased in other areas of Kadiri in the past five years. Approximately 78 trafficked women were identified in a single place. Some women were promised jobs as housemaids in distant cities but were sold to brothel houses in the metros mentioned above. Some women were sent without the knowledge of their parents. Using the abject poverty of the families, some men, on the pretext of love, have eloped and sold the girls to brothel houses. Lack of livelihoods and marriage difficulties (dowry) in such poverty stricken conditions have forced the girls to elope; often, falsely lured and trapped into the sex work net. Bhanuja lamented that the right to a decent livelihood is being eroded in the face of the severe challenges to survival.

The following tables shows the activities planned in terms of short term (less than a year) and long term activities (more than a year) drawn up by the RDRM/ ICPRD Workshop on Rayalaseema Regional Level Farmers Coalition Building on Combating Drought. Activities involving demonstration projects are to be taken up at least in one Gram Panchayat by each NGO while activities involving training, orientation, are being planned to be taken up in at least one mandal by each NGO. Resources for the activities and collaborations are being planned and sought.

Dry Land Agriculture

S. No.	Problem/issue	Activity	Strategy	Collaboration	Area	Duration
	Mono cropping	Shifting to other crops; Rotation of crops; intercropping Dry land horticulture: Demonstration Plots	Sensitising the farming community Distribution of seeds/saplings Seed banks Create common fund	Department of Agriculture, Horticulture, college of Agriculture; Banks	1 Mandal Village	Short Term
2	Usage of chemical fertilizers & Pesticides	Propagation and practices of Bio fertilisers, vermin compost, compost pits, integrated and non-pest management	Sensitisation/ Training to the farmers and exposure visits	Department of Agriculture; Banks	1 Village	Short term

3	Crop loss	Crop Insurance	Sensitising the farming community	Insurance Companies	1 Mandal	Short term
4	Economic usage of water	Drip/pot drip/sprinkler irrigation	Sensitising the farming community Facilitating loans from banks	Department of Agriculture, Horticulture; Banks	1 Mandal	Short Term
5	Spurious seeds Fertilisers	Monitoring for good seeds/fertilisers; purification process of seeds	Sensitising the farming community on the examination of seeds; on seed purification	Seeds corporations and companies	1 Mandal	Short term

Improvement of Natural Resource Management						
S. No.	**Problem /issue**	**Activity**	**Strategy**	**Collaboration**	**Area**	**Duration**
1	Water & soil Conservation	Desilting of existing tanks and ponds Strengthening of tank bunds Construction of check dams, soak its, New tanks Linkage of tanks	Identification of tanks and desilting Identification and strengthening through peoples participation Identification of areas and digging through peoples participation	DPAP, RWS Minor irrigation Grama Panchayats	1 Mandal	Long Term
2	Deforestation	Protection of forests through Van Sangrakshan Samiti Regeneration Top hill plantation	Seed dibbling/seed broadcasting/sowing Social fencing,	Grams Panchayats, Department of Forests, Revenue School children	1 Village	Long term

		for attracting rains Afforestration of Revenue lands & hillocks Usufruct rights to the people	Peripheral trenches Building people's Organisations			
3	Fodder Scarcity Drinking Water	Procurement of fodder from other places; Plantation of trees useful as fodder Raising fodder in tank beds/ waste lands; Minimising fodder wastage; Construction of water tanks for livestock near tube	Sensitising the farmers and VSS. Sensitising farmers on urea treatment and chopping fodder Lobbying	Animal Husbandry Forestry department Grama Panchayat MPDO	1 Mandal	Short term

Livelihood/ Migration / Trafficking /Convergent Community Action

S. No.	Problem/issue	Activity	Strategy	Collaboration	Area	Duration
1	Lack of food Security	Ensuring food Security	Rice Credit Line - Improvement of - Food for Work program - Annapurana - Antodaya as per Supreme Court order - Grain Bank Grain Storage bins	DPIP, DRDA District & State Administration AGRMI AFPRO	1 Mandal Do	Short term
2	Creating Alternative Livelihoods	Viable Income Generating activities Providing skills Micro- Finance delivery- SHGs (Financial Assistance) Marketing	Conducting studies (by an expert organization) for identification of viable income generating activities with market potential Mobilizing women into SHGs- Creating Access to	SC/ BC/ Minority Corporation / DWCW DPIP/ APRPRP / Drought Relief Commissioner ICPRD	1 Village	Short term

		Assistance	micro-credit Capacity Building			
3	Trafficking	Prevention	Building peoples coalition (including rescue)	Police/ Women's Groups Mytri Sanghams	1 Mandal	Short term
4	Stay . Shelter After rescue ./ return	Providing shelter and feeding	Providing shelter in shelter homes, short stay homes of Govt. / NGOs Facilitation for new homes		1 Mandal	Short term
5	Many similar activities being taken up individually by programs of different government department	Coordination between departments, projects of Government; Convergence	Lobbying with the district administration for a coordinated approach by the various departments and projects	District administration projects	1 Mandal	Short term

1 Mandal = Average of 70 – 100 Hamlets DPIP – District poverty initiative project; AFPRO – Action for Food program; DWCW – Department of Women & Child Welfare; APRLP – Andhra Pradesh Rural Poverty Reduction Project; MPDO – Mandal Parishad Development Officer; RWS – Rural Water Supply.

The threat of famine in Rayalaseema has attracted d countrywide attention though near famine conditions are endemic to this area. If the cry of distress has at last attracted attention it is because a continuous drought for five years has produced its deadly effects on a population whose economic resources and physical stamina have been progressively undermined by it. All the tanks and most of the wells, which arc the principal source s of water supply to the people, have completely dried up. Widespread stop page o fall agricultural operations, scarcity of food, high prices of food grains, and extensive unemployment among the landless agricultural workers, who constitute the over whelming bulk of the population all these physical manifestations of famine conditions are present.

The susceptibility of Rayalaseema to famine is not unknown to the authorities, and it is not as if the present calamity has com e suddenly and taken the authorities or the public unawares, On the contrary, that Rayalaseema is the epicentre of the famine belt of the Madras State has long been recognised and there ha s never been any dispute about it. In a monograph Rural Problems in Madras published by and under the imprimatur of the Madras Government a few years ago, it was admitted, for instance, that "the most insecure districts (from the point of view of liability to famine) are those known as the Deccan district, comprising of Kurnool, Bellary, Cuddapah and Anantapur, which have frequently suffered from famine during the last one hundred years" To these must be added the Chittoor district, adjacent to Madras, and forming for all practical purpose s an integral part of Rayalaseema.

The four districts mentioned cover a total area of 27,500 sq. miles with a population of about five millions, the vast majority of whom depend solely on agriculture for their livelihood. The average annual rainfall in the region is 25 inches, but it is only once in five years that Rayalaseema has an abundant monsoon instead of its being spread out evenly over the period, while in four years out of five, the rainfall is so precarious that cultivation is extremely difficult. The area is drained by the Krishna and the Pennar rivers and their tributaries, but there are practically no irrigation works in the area worth the name, with the exception of the Cuddapah Kurnool canal, the Kambam tank in Kurnool, and the Tungabhadra channels in

Bellary. The utilisation of the Tungabhadra waters to relieve the irrigational and drinking water problems of the area will be practicable only when the Tungabhadra project is completed and the canals distributing its waters begin to function. There are other irrigation project s under examination to utilise the Pennar and the Krishna waters for the benefit of the area, but non e ha s been taken on hand, let alone being in actual operation. As matters stand, in Rayalaseema at least, agriculture is a gamble in rain. The local government and its officials must take the blame for their inexcusable failure to appreciate the consequences of the drought during the preceding two years and for taking no heed of the warning provided by the failure of monsoon for four years. If at the first warning of the oncoming crisis, a properly planned scheme had been taken on hand for deepening wells, repairing minor irrigation works and constructing catchment basins for the mountain streams, and other measures essential for mitigating the effects of drought, the worst could have been averted. Nor would the crisis be so overwhelming in its incidence, if a systematic programme of road construction and of other emergency public works schemes had been kept ready for implementation, and put into operation as soon as the need for them became apparent. Had proper measures been taken in time, food production would not have been so poor, the economic conditions of the smaller ryots and landless labourers would not have become so deplorable as they are now and the word 'famine' would not have been writ large on the face of Rayalaseema.

Once the official machinery was logged into motion by the demands made upon it by the gathering volume of distress, relief measures have gone forward but even so, it is evident that they are not moving at the pace they should. Red tape takes its toll and relief is distributed in terms of the provision s of the Famine Code which, while "protecting the people from starvation in times of distress" does not impose much further obligations on the State.

Outstanding among the essential needs of the affected area is recognized to be provision of drinking water. Next in importance is employment for the agricultural classes, particularly landless labourers, whose resources have been completely exhausted on account of the absence of normal agricultural operations for the last three years. The third is the problem of preventing the outbreak of epidemics and diseases, incidental to prolonged malnutrition, undernourishment and the heat of summer and providing medical relief to the suffering.

Construction and repair of roads and repair of tanks and bunds have been the traditional methods of mitigating unemployment and ensuring necessary purchasing power of the poor labourer. According to the latest figure s supplied by Shri C. Rajagopalachari, Chief Minister, about a lakh of persons on the average are being employed in famine works. This does not, of course, exhaust the number of those who are without work but cannot get it and for whom employment has to be found. The number of such forced unemployed is variously estimated at about half a million. The problem of feeding the destitute and extending gratuitous relief to the disabled and the children is a complex and costly procedure 950 centres have been opened in the famine area at which nearly two lakh persons are being fed on standard gruel. Besides Government gruel centres, there are a few relief centres run by private agencies s whose number is unascertainable, where food is distributed, milk powder is supplied for children and so on. Even at the incredibly low cost of supplying the gruel, which works out at one anna per head per day, expenditure on this item of relief alone amounts to Rs 14,000 per day, nevertheless, even under the Famine Code, financial considerations are inapplicable when the desideratum is prevention of starvation. In the case of a number of persons, the gruel supplied at the feeding centres is a supplementary ration, adding to the food which can be procured at the fair price food grain shops. On the other hand, from the standpoint of the people in distress, the price of rice, millets and ragi at these shoos is prohibitive, and a proposal to sell millets, the staple food grain of the rural population, at subsidised rates, is still under examination by the Government. The Army has been pressed into service for deepening wells and relieving water scarcity. Under the supervision of Lt. Col Talwar and Lt. Col Mullenix, Army Engineers, the Army has been engaged in the stupendous and urgent task of fighting the famine on the water front both by deepening existing wells where they had gone dry and by utilizing Army trucks and water curriers to distribute drinking water to the villagers in the interior, where its scarcity is most acute. The extreme scarcity of water in this area renders the services of the Army doubly welcome The speed at which the well-digging is going on may not be as rapid as one would wish, but the Army is proceeding according to a prescribed programme, beginning with Cuddapah district and extending its operations to Kurnool and then to Anantapur districts, and results cannot be striking when, as has been found by experience, some wells had to be deepened up to 60 feet or more, before water is struck. It is good to know, however, that consideration of cost is not allowed to deter efforts in this direction.

Movement of food grains into the affected areas is also not as expeditious as it should be or could be but in any case, unless the problems of high prices and absence of purchasing power in the hands of the vast majority of the working population are resolved by dispatching larger quantities of millets and ragi to the affected areas and the fair price shops are well-stocked with grains while more repair and construction work is undertaken which would provide employment to the able-bodied, the pressure on State resources and public charity will remain undiminished. The long-range social effects of gratuitous relief will also be highly deleterious.

The part played by private and philanthropic organisations in providing relief, however, deserves every commendation. The State is concerning itself mainly with providing grants for well-digging and land improvement and in making subsistence grants to enable the smaller ryots and the middle class families to get through the crisis that confronts them. While the brunt of the burden of providing relief is being borne by the State Government, to whom the cost cannot matter so long as distress persists, the Government of India have made available a sum of Rs 2.48 crores for subsistence loans and financing of gruel centres. Other sources for relief operations arc the Prime Minister's Relief Fund, from which a grant of Rs 3.75 lakhs has so far been made, and the Governor's Relief Fund, contributions to which total Rs 2½ lakhs now. The Rayalaseema Relief Committee of the Andhra Pradesh Congress has collected about Rs 2 lakhs, while the "Express" group of papers fund has swelled to about Rs 2 lakhs. Besides, there has been a liberal flow of private charity from the surplus northern Circars districts and some districts in the South in the shape of grain and clothes and, what is equally important food for the cattle. The cattle have been as unfortunate and helpless victims of the water and fodder famine in Rayalaseema as men, and feeding and keeping them alive through the current distress period is a vital need for the agricultural economy of the area. Among the organisations rendering notable service by managing gruel centres, distributing cloth and organizing other kinds of relief, mention must be made of the Ramakrishna Mission.

If action to relive distress is delayed till the crisis actually overwhelms the people, it involves expending much larger sums of money besides, of course, much avoidable misery, than if preventive measures are taken well in time. Even as far back as 1938, the Madras Famine Code Revision Committee prescribed the remedial measures essential in such conditions, State intervention has up to the present been largely designed to save life and mitigate suffering.

Among others, the Committee recommended the following multi-directional programme, development of irrigation, namely, provision of irrigation facilities and tank restoration, introduction of improved methods of agriculture, such as bunding and dry farming, exploitation of the mineral resources and development of large scale industries and cottage industries. These remedies are as appropriate and necessary today as they were in 1938, though not quite adequate in themselves to meet the demands of the situation that has arisen. In the intervening years, precious little has been done to give them a purposeful and vigorous push. What Rayalaseema needs more than anything else is protected irrigation, as an insurance against the ravages of drought and as a means of increasing food production and development of industries, big and small, for which her rich natural resources, so far untapped, afford abundant opportunities and scope. The Tungabhadra project, now under construction, is a part of the solution, but it is neither an immediate nor a full solution, because, the project will take another two years to complete and its benefits will be derived largely only by the Bellary and parts of Anantapur districts. Other irrigation schemes are essential, besides the execution of an extensive programme of minor irrigation works and construction of wells, which can be worked by pumps, wherever possible, run on electrical energy. The Rayalaseema Development

Board, set up in 1941, solely to carry out schemes which would contribute directly to the prosperity of the area, has no spectacular achievements to its credit. The financial resources at the disposal of the board are meagre, not commensurate with the requirements of the situation and a number of its recommendations have got enmeshed in departmental red tape. If it is converted into a statutory corporation and sufficient funds are placed at its disposal, and if it is enabled to initiate and execute development schemes on its own responsibility, it can prove an effective agency for good.

Shri C. Rajagopalachari, Chief Minister of Madras, has said that the famine conditions in Rayalaseema might ease by September, by which time the area would have received the benefit of the main monsoon. That is, of course, dependent on nature and providence, but in the meantime there can be no relaxation in the relief operations or in efforts to devise and push through long-range schemes for famine prevention. The farmers can turn to the land and make it yield its fruit only if sufficient water is made available to them.

In this direction, it is only purposeful and systematic State action that can prove helpful. It is also essential that some of the irrigation and power schemes, which would directly benefit this backward area, should be taken in hand under the Five Year Plan and the Planning Commission has a special responsibility in this regard. Famine conditions, though not as acute as in the five Rayalaseema districts, obtain also in some other farts of the Madras State in parts of Guntur and Nellore districts, in South Arcot, Ramnad and Chingleput districts. This fact emphasise that about a third of the Madras State experiences conditions ranging from shortage to scarcity year after year. This year, the rice production in the State is in deficit to the extent of a million tons and production of millets by about four lakh tons. This deficit can be made good mainly by the Centre, which lias already declared that it would supply only about half the quantity, of which no more than two lakh tons will be in the shape of rice. Fortunately, the surplus districts had a good crop, and to them the Government must look for fulfilling the procurement targets, which alone can obviate a greater shortage later in the year. The increase in procurement prices for paddy announced by the Government should prove an inducement, but to counterbalance it, there is the increase in food prices consequent on the stoppage of food subsidies by the Government of India. Altogether, what with famine and unforeseen expenditure on relief measures on the one hand, and the potential danger of food shortage, which may grow as the year advances, on the other, both the finances and the general economy of the State as a whole face prospects which are dismal indeed.

Important Political and Economic Changes:1854-1866

Between the famines of 1854 and 1866 certain important political and economic changes took place in India. The effects of which were experienced by Andhra along with other parts of the country, Politically, India witnessed a great rebellion in 1857, which shook the British raj to its foundations. The British, however were able to reassert their authority after suppressing the movement with great effort. The East India Company's rule was abolished. The Crown and the Parliament assumed direct responsibility for the Government of India in 1858. The administrative machinery was geared to buttress their rule and during the next half century paternalism became the keynote of their policy.[44] India was made the "pivot of the British Empire.

From the economic point of view. India came under the vortex of the British industrial capitalism. By the middle of the nineteenth century Britain made rapid strides in Industrialisation. Steam power, steel, cement, machine tools, etc. began to play a dominant role in British Industry. It became the "forge of the world, the world's carrier, the world's banker, the world's workshop, the world's clearing house and the world's entrepot".[45] The emergence of France, Germany, U.S.A. and even Japan on the industrial map of the world created a competition among the industrial powers including Great Britain for the control of raw materials and markets. Such a control was deemed essential for the very survival of these advanced countries. In this context Britain regarded India as a "prized possession" not to be renounced under any circumstances.[46] Viewed from this angle, it is no surprise that the British saw with skepticism attempts by enterprising Indians in the latter half of the nineteenth century to mechanise their industries. Many direct and indirect steps were taken to prevent Industrialisation of India almost up to the end of the British Raj.

Development of Communications

To facilitate the movement of the British goods into the interior parts of the country and to fetch raw materials and foodstuffs, railroads, metal roads and navigable canals were constructed from the middle of the last century. Whereas in 1853 there were only 20 miles of railway in India, by 1866 it rose to about 4,000 miles, and by the end of the century to 25,000 miles built at a cost of Rs. 340 crores.[47] These railways did not help to promote the Industrialisation of India. Agriculture, therefore, continued to be the main occupation of the people.

As for Andhra most of its towns were connected by rail and road with the rest of the Madras Presidency by 1866. The towns of Nellore, Guntur, Vijayawada, Machilipatnam, Visakhapatnam, etc. were linked with the interior parts of the region and with places like Madras, Bangalore, Hyderabad and Bombay Among the towns in Rayalaseema which had communication facilities in the sixties of the last century mention may be made of Anantapur, Bellary, Cuddapah, Kurnool, Adoni, Guntakal, Chittoor, Tirupati, etc.[48] The availability of such outlets for farm produce contributed to the growth in the export trade and to the general rise in prices.

Another cause for the increase in prices in Andhra and elsewhere during the sixties and the seventies of the last century was inflation due, among others, to the steady growth in the population [49] and the consequent increase in the demand for the necessities of life. The population of Andhra (including Rayalaseema) rose from about 9.8 million to about 12.5 million between 1855 and 1865.[50] In the next decade it rose to about 19 million. As for Rayalaseema, the increase was from about 4.4 million to 4.9 million between 1855 and 1865 and to 6.4 million in the succeeding decade. Further the influx of European capital into India in the form of investments in railroads, irrigation works, navigation canals, plantations, etc. from the mid-fifties increased money supply in the country. The adoption of cash crop culture in preference to food crops added to the dearness of food. The ryots of Rayalaseema, as their brethren in the rest of the country, devoted their "best and most fertile land"[51] to the cultivation of cotton in view of the handsome prices obtained following the outbreak of the civil war in the U.S.A. (1861-65). The home authorities had also pressurized the Government of India to encourage cotton cultivation to enable the British textile industry to tide over the cotton crisis.[52] The acreage under cotton had thus more than doubled in Bellary (including Anantapur), Cuddapah, North Arcot, Kurnool, Ganjam and Visakhapatnam between 1862 and 1864.[53] The average price of cotton in Rayalaseema rose from Rs. 70 in 1861 to Rs. 240 in 1864 per candy of 500 lbs.[54]

The net result of all these factors was the unprecedented rise in the price of food grains in Rayalaseema and elsewhere in Andhra. The increase in the prices ranged from 50 to 100 percent[55] in Circar districts of Ganjam, Vishakhapatnam, Godavari and Krishna between 1855 and 1865.[56] The same was the case in Nellore and Rayalaseema.[57] The following tables would give a comparative picture of the prices of food grains in the several districts of Rayalaseema since 1859 and would illustrate the gravity of the problem in the famine year of 1866.[58] The increase in prices naturally affected the vulnerable sections of the society whose sources of income continued to be meagre.

Table 3 Food grains of Rayalaseema 1859-66

I Average prices of Rice, etc. in the Bellary district in rupees per Madras Grass of 3200 measures:

Items	Year					Feb. 1866	Sep. 1866
	1859-60	1860-61	1862-63	1863-64	1864-65		
Second sort Rice	310	326	371	477	567	355	355
Second sort Paddy	139	147	178	200	229	342	342
Cholam	193	167	182	317	208	603	793
Ragi	167	164	160	266	260	573	763
Horsegram	186	178	184	309	342	687	922

II. Average prices of Rice, etc. in the Kurnool district in rupees per Madras Grass of 3200 measures.

Items	Year					Feb.1866	Sep.1866
	1859-60	1860-61	1862-63	1863-64	1864-65		
Second sort Rice	341	345	362	472	483	611	706
Second sort Paddy	148	150	156	206	239	292	349
Cholam	171	176	184	306	258	437	540
Ragi	154	165	167	257	256	391	461
Horsegram	191	208	207	291	307	504	616

III. Average prices of Rice, etc. in the Cuddapah district in rupees per Madras Grass of 3200 measures.

Items	Year					Feb.1866	Sep.1866
	1859-60	1860-61	1862-63	1863-64	1864-65		
Second sort Rice	334	356	430	427	408	603	664
Second sort Paddy	149	162	200	200	230	275	313
Cholam	195	215	21	261	258	403	484
Ragi	184	204	200	230	242	370	460
Horsegram	212	225	213	257	264	454	577

IV. Average prices of Rice, etc. in the North Arcot district in rupees per Madras Grass of 3200 measures

Items	Year					Feb. 1866	Sep. 1866
	1859-60	1860-61	1862-63	1863-64	1864-65		
Second sort Rice	256	303	350	323	341	513	614
Second sort Paddy	107	144	157	145	158	224	281
Cholam	116	158	231	194	227	356	441
Ragi	155	184	201	175	189	337	413
Horsegram	192	223	194	177	211	447	570

Excessive Taxation

The increase in the price of food did not contribute to the prosperity of the agriculturists owing to the high incidence of taxation. In the ryotwari areas of Rayalaseema and elsewhere in Andhra land tax continued to be heavy even after 1857.[59] When the new revenue settlement was put, into operation it was generally, agreed that land revenue should not exceed one half of the net produce.

The secretary of State Sir Charles Wood, reiterated this point in his dispatch of 1864.[60] However, as Sir Louis Mallet has rightly observed, "In practice the rates levied have often absorbed the whole rental and not infrequently. I suspect encroached on profits also."[61]

Besides land revenue and water assessment (amounting to over 50 percent of the net produce) payable to the Imperial Government, the ryots in Rayalaseema and in such other ryotwari areas were obliged to pay additional sums to local Government in the form of cesses. Impartial observers, therefore, concede the fact that, by and large agriculture taxation "proved to be excessive in the Madras Presidency" even in the third quarter of the nineteenth century. There is an interesting passage in an exchange of notes between the Boards of Revenue of the Madras and the Bengal Presidencies with reference to the incidence of land revenue in these two provinces. G.S. Forbes, Collector of Ganjam, in his remarks on the Bengal's proposal for increasing the tax on salt in Madras, observed, when the Bengal authorities ask why the inhabitants of Madras districts should pay less for their salt than those of Bengal, it might justly be retorted on them, why should the Bengal ryot pay a lower land tax by about forty percent than the Madras ryot.[62]

All this would show how the ryot of Rayalaseema was not only overtaxed but was also not certain of how much he should pay next year in view of the increasing demands of the Government. Until 1880, even the 30 years rule for the revision of assessments was not adopted causing great hardship to the cultivators in the Ryotwari areas of the Madras presidency.

Their counterparts in the delta districts felt the pressure of land revenue more by the ryots of Rayalaseema. Whereas the ryot of Rayalaseema paid around Rs. 12 as land revenue for a holding of

10 acres of which about 90 percent constituted dry land, the ryot of Godavari paid Rs. 19 for a holding of 11 acres of which 61 percent was classified as wet.[63] As the latter was able to produce more grain, he was able to obtain better profit than the former and was able to pay the land revenue without much difficulty. [64] On the other hand, the Rayalaseema root had to devote greater energy and resources to produce grain from a generally inferior soil practically with no assured water supply.[65] For want of resources majority of these roots were unable to effect any improvements in their land.[66] Therefore, except in the years of normal rainfall. they could not afford to pay the land revenue promptly and fell into arrears. The rigidity of the rules of remission added to their woes. Moreover, land had very little salable value, particularly in the districts of Bella and Anantapur. Frederick Nicholson, Collector of Anantapur, vouched this fact in his report of 1887, he observed "the ryot [of these district] has little exchange value in land, it is hardly a marketable article, no loan could be had by mortgage of lands."[67]

This inevitably led them to seek loans from the moneylenders who generally belonged to the trading community. They advanced money at usurious rate of interest ranging from 24 to 50 percent, which was collected with the principle at the time of harvest.[68] To clear the loan the roots were obliged to part with their produce at a price far below the market rate. After discharging all their obligations they were left hardly with any surplus to tide over the loan seasons. They therefore, fell victim to the vicissitudes of drought and famine more easily than their counterparts in the coastal districts of Andhra.

Lack of irrigation facilities made the roots of Rayalaseema depend mostly upon monsoon rains, which was normally below that of the coastal districts. While the average annual rainfall in coastal Andhra exceeded 40 inches, it was generally below 30 inches in Rayalaseema. Among the districts of the region, Bellary (including Anantapur) happened to be the driest. In the five years preceding the famine of 1866 the annual rainfall in the district was far below the average of 23 inches. As a consequence it became the theatre of famine in that year, paradoxically, Ganjam, which happened to be one of the wettest districts of Andhra, receiving an average of annual rainfall of over 45 inches, fell victim to the famine of 1866 due to the unprecedented drought of 1885-66. The construction of the K.C. Canal in the sixties of the last century provided to sonic extent irrigation facilities in Kurnool and Cuddapah districts.[69] Usually, therefore, the effect of drought was not as severe in these districts as in Bellary. However, the extreme distress suffered by the people of Kurnool and Cuddapah during 1876-78 famine along with those of Bellary, North Arcot, etc. revealed the fact that they too were not immune to famines caused by severe droughts.

Condition of Landless Labourers

The condition of landless labourers, the bulk of whom was drawn from the lower castes of the society such as Malas, Madigas and Boyas continued to be deplorable in Rayalaseema during the second half of the nineteenth century.[70] The relative poverty of the ryots prevented them from engaging laborers in the same measure as their counterparts in the delta districts.' The majority of the labourers were attached to substantial ryots. Their wages usually were paid in kind which ranged from about 1½ Madras measures of grain to 2 per day during the agricultural seasons.[71] In some places cooked food was also given as well as particularly during busy seasons of sowing and harvest.[72] In the absence of alternative sources of employment, their wages remained pretty low. During the off seasons when there was no work, the labourers fell easy prey to the money lenders, merchants and well-to-do ryots who lent money in advance and extracted 'kork cheaply during the cultivation seasons'.[73] So their living conditions continued to remain as miserable as ever. The Collector of Kurnool observed, their food and clothing were of the same kind as what they were before. They ate the same coarse grain and used as condiments the same chutney composed of hemp leaves or tamarind fruit. They wore the same coarse clothes and slept on the same rope cots.[74]

The labourers in Kurnool and Cuddapah, however, were somewhat better off as they could Lind works on the K.C. Canal project, where an ordinary labourer working with his wife and children, was paid between 12 annas to one rupee per day.[75] The Canal company, however could not provide employment to all those in need of alternate work when agricultural operations retarded due to drought.

In the North Arcot district, where the practice of cash wages was prevalent the wages did not increase commensurate with the increase in prices.

During 1860's male coolie was paid 2 annas and a female coolie 1 anna per day during the cultivation season. When rains failed even this small wage could not be obtained while high prices kept food beyond their reach. Hence they became the first victims of famine, skilled workers, such as carpenters, masons, etc. earned nearly a rupee per day as wage. They also became vulnerable to the effects of famine since their principal customers were the agriculturists.

Low-paid Government Servants

The condition of the low-paid Government employees was deplorable during the sixties of that last century. A peon in the revenue office, for instance, received a monthly salary between Rs.3 and 7 and a lower division clerk received about Rs. 15. These wages had not undergone any significant change ever since the establishment of the British rule in India. In consequence, the family of a low-paid employee survived on hardly one meal a day. Requesting for an amelioration of their lot, J.I. Minchin, Collector of Kurnool, wrote to the Board of Revenue: "I am sure it will meet the full consideration of Government and that measures will be taken to relieve the low paid servants of the state from their present difficulties and distress".[76] Nothing seems to have been done until the great famine of 1876-78, when the Government sanctioned an increase of Rs. 2 to the salaries of its native employees.[77]

The Great Famine of 1876-78

In the annals of Anantapur, the years 1876-78, corresponding to the Hindu cyclic years *Dhatu* and *Iswara* have attained great notoriety. For, during those two years this region was visited by a famine which had truly "put all the subsequent, as it did all the previous afflictions in the shade."[78] It had infact affected vast tracts of India south of the Vindhyas. In the Madras Presidency 14 out of the 21 districts experienced famine, of which the most affected were the districts of Bellary (including Anantapur), Kurnool, Cuddapah, Nellore, North Arcot, Salem, Madura, Chingleput and Coimbatore.[79]

Almost all sections of the society barring the traders, were affected by the famine.[80] The small ryots, the landless agricultural labourers, the artisans and the low-paid employees suffered most from the rigour of famine as on earlier occasions.

To sum up the narrative of this chapter, the scarcities that India experienced during the sixteen years, 1880 to1895, were of local nature and confined to such limited area that they did not make any difference to the overall supply of food grains in the country. The rise in the prices in each case was consequently limited and food was freely available to the people in the market. The problem in each case 85 was, therefore, more or less of providing work to the unemployed to enable them to earn wages to purchase food.

Notes and References

1. Dalyell, *Report*, p. 107 quoted in Famine Histories p. 74 this writer of Famine Histories, however, points out that "this seems a forced concession from Ellis's report which is to the effect that though prices might be his at home, they would seed grain away to other places where it was higher raise than raise prices to the same high rate at home."
2. *Ibid.*
3. Daylell underestimated mortality at 2 lakhs. The figure of 4,50,000 is that of the Famine Commission of 1880. Vide *Famine Histories,* p. 74.
4. Dalyell, *op. cit.*
5. Vide Famine Proceedings Nos. 19 and 23, containing of Reports from Madras dated 12th January and 23rd January 1878 respectively to India.
6. Letter from Madras Government to government of India, 30 November 1876. Vide Famine Proceedings No. 51, December 1876.
7. Letter from India to Richard Temple No. 198, dated 16 February
8. Cornish's reply to Richard Temple on reduction of wages, video proceedings No. 10, May 1877.
9. Telegram No. 43 (Revenue), 3rd May 1877.

10. *Madras Famine Review*, Appendix 'P', quoted in Famine Histories.
11. *Ibid.*, p. 211.
12. *Ibid.*, p. 255.
13. *Ibid.*, p. 264.
14. Frederick Goodall, 'Famine in Mysore', *Calcutta Review*, vol. LXXXVI, pp. 261-76.
15. Vide Memorandum concerning the scarcity in the Nizam's territory Famine proceedings No. 113, March 1877.
16. *Famine Commission (1898) Report*, Chapter II, para 20.
17. *Ibid.*, para 21.
18. *Ibid.*
19. This was said to be average over the preceding 20 years, vide *Attrition Report, Central Provinces, 1886-87*, p. 29.
20. Madras to India, No. 226 (Rev.), dated 22 March 1892, statement "A".
21. Report of Madras famines, Madras to India, dated 6th September 1893, vide famine proceedings May 1894.
22. Famine Commission (1898) Report, Chapter II, para 36.
23. Madras Government letter to Secretary of State, 21st June 1892.
24. The appellation was later given to the famine of 1899-1900.
25. Later called United Provinces of Agra and Oudh.
26. Holderness, Narrative of the Famine in India in 1896-97 (vide Government of India R & A Department Famine Proceedings, December 1897). Also printed separately in a book form (Simla, 1897).
27. Famine Commission Report 1898, p. 235.
28. Holderness, *op. cit.*, para 27.
29. Vide note of Bombay Government Revenue Department, 336 (famine date 9th May 1898).
30. Vide report on Famine in Madras para 43, Madras Revenue and Agricultural Department proceedings, July 1898.

31. Vide Letter from Bengal to India No.7-TR, dated 3rd November 1896.
32. India to Bengal dated 4 January 1897.
33. Resolution of Government of India No. 33 of 1897, dated 18th May 1897.
34. Bengal to India No. 5183, dated 12th December 1896 and also No. 25, dated January 1897.
35. Bombay Government Resolution No. 9296, dated 21st November 1896.
36. Holderness, *op. cit.*, para 21.
37. Report of Irrigation Commission, part I, para 92.
38. Despatch from India to Secretary of State No. 18, dated 18th May 1986.
39. *Ibid.*
40. Resolution No. 18, Civil Works of 1905 (irrigation), dated 18th May 1905.
41. Despatch to secretary of State No. 18, dated 18th May 1905, para 4.
42. Lady hope, Life and Works of General Sir Henry Cotton, Chapter XIII.
43. Summary of Lord Curzon's Administration R and A Department, part I, para 22 to 52.
44. W.R. Cornish. *Report of the Census of the Madras Presidency 1871,* Vol. 1, pp. 49-50.
45. Lord: Curzon spoke of India's central position in the *Pax Britannica* and firmly believed that if India slips out of the British hands, then "our Should sink to its Setting", quoted in Durga Das, *India from Curzon to Nehru and After* (London, 1969), pp. 44-45. Disraeli, the Prime Minister of Great Britain during 1870s called India the "brightest jewel", Crown of Great Britain - Michael Edwardes. *The Last Years of British Rule in India,* London, 1963, p. 1.
46. L.C.A. Knowles, *The Industrial and Commercial Revolutions in Great Britain during the Nineteenth Century,* London, Reprint, 1950, p. 139.

47. Between 1862 and 1882, the import duties on the finished products of Britain were abolished under pressure from the industrial lobby in Britain. When the Government was in dire need of money it was compelled to reimpose the tariff duties on British goods in 1894. However, care was taken to exempt the Lancashire and Manchester textiles from these duties. In order to protect the British goods a counter-wailing excise duty was levied on Indian manufactures. These 'unjust-duties' were not abolished until 1925. In 1932, India was obliged to sign the Ottawa Trade Agreement, which gave preferential treatment to British imports into this country. Indian commodities exported to the countries outside the British Empire were taxed heavily. This step-motherly attitude to Indian industries continued up to the independence of India. Romesh Dutt, *The Economic History of India in the Victorian Age* (London, Eighth Edition, 1956) pp. 400-413, 536-544. See also Manoranjan Jha, *Role of the Central Legislature in the Freedom Struggle* (New Delhi, 1972), pp. 10, 193-196; V.V. Bhatt, *Aspects of Economic Change and Policy in India 1800-1960* (Bombay, 1963), pp. 46-50.

48. Romesh Dutt, *op. cit.*, 1956, pp. 548-49.

49. Bipan Chandra, *The Rise and Growth of Economic Nationalism in India*, New Delhi, 1969, p. 181.

50. C.D. Malceane, *Standing Information Regarding the Administration of the Madras Presidency*, Madras, 1879, p. 298. See also Quarter Master-General of the Madras Army, *Routes in the Madras Presidency*, Madras, 1879, pp. 1-4, 23-25, 30-32, 94-95, 185-191 and 378-379.

51. W.R. Cornish, *Report on the Census of 1871*, Madras, 1874, p. 9. See also Lewis McIver, *The Imperial Census of 1861 — Madras Presidency*, Madras, 1883, p. 85.

52. P.J. Thomas and B. Natarajan, *Economic Depression in the Madras Presidency, 1820-1854*, Madras, 1933, pp. 16-17.

53. J.I. Minchin, *Collector of Bellary to the Board of Revenue*, Madras, 12th May 1864. *Proceedings of the Board of Revenue*, 10th August 1864 (A.P.S.A.).

54. *Revenue Consultations*, Madras, 4th August 1861, (A.P.S.A.), *Proceedings of the Board of Revenue*, Madras, 18th June 1861 and 9th August 1864 (A.P.S.A.).

55. *Proceedings of the Board of Revenue, op. cit.*

56. R.A. Dalyell, *Memorandum on the Madras Famine of 1866*, Madras, 1867, p. 62.

57. *Ibid.* The average price of second sort rice rose from Rs. 116 per tonne of 3200 Madras in 1855 to Rs. 230 in 1865 in Ganjam. In the other three Districts the increase was from Rs. 166 to Rs. 260.

58. *Ibid.* In Nellore the price of second sort rice rose from Rs. 182 to Rs. 300 between 1855 and 1865. In the districts of Bellary, Kurnool and North Arcot the average price of second sort rice increased from Rs. 194 to 340 in the same period.

59. *Proceedings of the Board of Revenue*, Madras, 18th June 1861, 9th August 1864, 24th February 1865, 21st March 1866 and 22nd October 1866 (A.P.S.A.).

60. Although the ryotwari system was intended to advance the interests of the ryots by its innovators, "it did not fulfill the promise held out by them". As Nilamani Mukherji aptly put it "the chief arms of the new mode of land revenue administration had been good collection of revenue and amelioration of the condition of the ryots. The system realized the first objective admirably well, in the second, it failed: Nilamani Mukherji, "The impact of the Ryotwari System on the Economy of the Madras Presidency." In B.N. Ganguli (ed.), *Readings in Indian Economic History*, Delhi, 1964, pp. 135-136.

61. *Revenue Despatch to Madras*. 24th February 1864, No. 7 in Romesh Dutt, *The Economic History of India in the Victorian Age*.

62. Eric Stokes, *The English Utilitarian and India*, Oxford, 1959, p. 109.

63. Romesh Dun, *op. cit.*, p. 340.

64. S. Srinivasa Raghavaiyangar, *Memorandum on the Progress of the Madras Residency*, Appendix, Section V, p. CC1.

65. It is estimated that during the sixties of the last century, an acre of dry land in the Presidency yielded about 190 seers of rice as against 370 seers obtained from an acre of wet land. See R.A. Dalyell, *Memorandum on the Madras Famine of 1866*, p. 70.

66. Although the average holding of 15 acres of ryotwari land in Bellary was the highest in the Presidency still this happened to be the most backward district in the province.

67. S. Srinivasa Raghavaiyangar, *op. cit.,* Section V, p. CC1.

68. Quoted in W. Francis, *Anantapur District Gazetteer*, pp. 53-54.

69. The average rainfall during this period was 18 inches. In 1865 only 8 inches of rain fell in Bellary. R.A. Dalyell, *op. cit.*, p. 51.

70. The K.C.Canal having a length of 190 miles diverts part of the Tungabhadra water to irrigation in Kurnool and Cuddapah. It was built by a British owned Company and was nationalized in 1882.

71. W.W. Hunter, *The Imperial Gazetteer of India*, Vol. 111, New Delhi, Reprint, 1975, pp. 328-329.

72. *Report of the Indian Famine Commission*, 1898, Vol. 1, pp. 360-361.

73. S. Srinivasa Raghavaiyangar. *op. cit.*, Section III, pp. IXXX, VII.

74. *Indian Famine Commission 1898*, Appendices. Evidence of Witnesses from the Madras Presidency taken before the Indian Famine Commission, 1898, pp. 24-25, *Proceedings of the Board of Revenue*, 13th October, 1876, No. 2576.

75. *Proceedings of the Board of Revenue*, Madras, 11 November, 1872, quoted in S. Srinivasa Raghavaiyanuar, *op. cit.*, p. LXXXI.

76. *Ibid.*, J.D. Robinson to the Board of Revenue, 12th May 1864, 10th August, 1864.

77. *Ibid.*, J.I. Minchin to the Board of Revenue, 12th May 1864.

78. Romesh Dutt, *op. cit.,* p. 363.

79. W. Francis, *op. cit.*, p. 90.

80. *Proceedings of the Board of Revenue*, 7th September and 24th October 1876.

Chapter V

Famines and Agrarian Distress in Anantapur District

The problem of famines and food shortages is one of the most acute facing agricultural economists. Since the mid 1970s Amartya Sen's approach to the economics of famine has become influential. He has argued at some length that a major cause of famines is not a sudden decline in food availability, but a sudden redistribution of what food is available. It will be argued here that there are major weaknesses in his theory which mean that it is more likely to cause famines than to cure them. It will be argued that his theory and analysis are wrong and that there are inconsistencies between the arguments he presents. Furthermore, the implications of his theory are contradicted by the facts given in the sources he uses. There are also ubiquitous and systematic inconsistencies between the facts he gives and the facts given in his sources. This monograph will concentrate on Sen's analysis of the Bengal famine of 1943, as it is the one he gives most attention to, it is the best-documented one, and it is the one for which his theory is most plausible. To be absolutely fair to him, the analysis will rely entirely on the sources he quotes and no new evidence will be presented.

Both Amartya Sen and the economists who worked in the mainstream tradition accept that famine can arise in one of three ways. The first is when there is a fall in the aggregate food supply of a country or region. This may happen because of drought, flood, plant disease (the potato blight), pests (the plague of locusts), blockade, post-harvest losses, or disruption of production by migration, epidemic or war. Problems with storage or transport or the manipulation of stocks by speculators can mean periods of famine even in a year in which there are adequate supplies.

The second way is an increase in aggregate demand. This may arise from immigration, an influx of refugees or an invading army. It has happened that foreign buyers or an occupying army have been able to export the food supply. A change in farmers' willingness to sell may be best treated under this heading, as a change in reservation demand.

Third, a change in the distribution of what food is available between people in different occupations may bring about a famine when there is no food shortage. For example, hyper-inflation, unemployment or enclosures have left some people without the means to buy food. Changes in income distribution may mean that some people eat more, leaving less for the others. These changes should not be confused with changes in geographical distribution because of transport failures, etc. which would come under a fall in supply. Nor should these changes be confused with the maldistribution that exists to a greater or lesser extent in every society, which may cause chronic malnutrition among thc poor, but which does not cause famine. It is a truism that the world could feed itself, so all famines are due to a change in relative distribution, if the famine area is defined sufficiently widely, to do this ignores reality, there are political, social, physical and economic constraints to redistribution which cannot be assumed away.

It is clear that, recurrent droughts in the region had their impact from time to time, the resultant responses or adaptations that have been taking place are discussed here. Before discussing the different adaptations to drought, it should be noted that responses vary from one specific drought year to another. Further, these adaptations may not take place in any particular sequence nor does each jati follow any one set-pattern.

Anantapur is the southern-most district of the Rayalaseema region of Andhra Pradesh. While agriculture remains the most important economic activity of the district, it is characterised by high levels of instability and uncertainty. Being located in the rain-shadow region of Andhra Pradesh, the district is drought-prone. By now, it is well established that while a generalised rural crisis is prevalent across the country, a disproportionate burden has fallen on its drier tracts. In 2006, Anantapur was one of the thirty-one districts identified by the Government of India as being prone to agriculture-related suicides. A few committees have been set up by the Government of Andhra Pradesh to study the agricultural crisis as well as the problem of farmers' suicides in the state. While these committees have looked at a whole gamut of issues underlying the agricultural crisis of Andhra Pradesh, our study of Anantapur district focuses specifically on the technological dimension of the crisis. Our study addresses such factors that have a bearing on agricultural production and productivity, namely, the physical environment, the extent and nature of agricultural inputs used, the nature of crop protection practices followed, the overall management practices adopted in cultivation including irrigation practices and aspects of technology delivery.

Anantapur is the largest district in Andhra Pradesh, its area measures 19130 sq km, this is almost equivalent to the district of East and West Godavari district area of Andhra Pradesh both these district is mere 32 lakhs. This is an indicator of its development. Where there is no opportunity of development the population is naturally thin. Agriculture land of Anantapur could not attract anyone to migrate over here. On the other hand the district has history of migrations. In the last 40 to 50 years landless Dalit and poor have left and settled in other places. Here 90% of the cultivable land is depending on the rain fed, even this rain is erratic and it does not guaranty any crop.

It is only agriculture, which is main source of employment in the district even to this day. Lack of rainfall has forced the farmers to go in search of ground water and this has forced the farmers to depend on power. A modern cultivation method, wide spread usage of chemical fertilizers and pesticides has made agriculture for more cost oriented. This has transformed from completely labor oriented agriculture in to market oriented and power dependent agriculture. The fluctuations in the market have played havoc with the lives of the farmers. So, not in a position to cope up with the agriculture returns, every year former is forced to go for more and more debts. It has reached a state that the farmer is forced to go for cultivation even in the face of recurring losses. So, in these conditions of scanty rainfall and marketability of Groundnut invariably the only crop turned out to be the Groundnut, which could satisfy the farmer's demands and also government policies.

Keeping previous chapters as the background, we probe into the details of the study and analysis of various facets of famine and drought, to which Anantapur district fell as a victim right from the historic period (information is available only from fourth century CE) to the present day.

Historical Background

Anantapur District was formed in the year 1882 having been separated from Bellary District. Later on it was expanded with the addition of Revenue Mandals of Kadiri, Mudigubba, Nallamada, Nambulapulakunta, Talupula, Nallacheruvu, Obuladevaracheruvu, Tanakal, Amadagur, Gandlapenta (previous Kadiri Taluk) from Cuddapah district, in the year 1910. During the year 1956, the present Revenue mandals of Rayadurg D. Hirehal, Kanekal, Bommanahal and Gummagatta of Bellary district were added to Anantapur district. The District has been divided into 3 Revenue Divisions consisting of 63 Revenue Mandals (Anantapur Division 20, Dharmavaram 17 and Penukonda Division 26).

Boundaries and Topography

Anantapur District lies between 13^{0}40' and 15^{0}15' North Latitude and 70^0 and 78^{0}30' East Longitude. It is bounded by Bellary, Kurnool districts on the North Cuddapah and Kolar districts of Karnataka on southeast and north, respectively. The district is roughly oblong in shape the longer side-running north to south with a portion of Chitradurg district Karnataka State intruding into it form west between Kundurpi and Amarapuram Mandals.

The district may be divided into 3 natural divisions. They are:

1. Northern Mandals of Rayadurg, Kanekal, Beluguppa, Gooty, Guntakal, Vajrakarur, Uravakonda, Vidapanakal, Yadiki, Tadipatri, Putlur and Yellanur containing large areas of Black cotton soils.

2. Kalyandurg, Kambadur, Settur, Brahmasamudram, Ramagiri, Kanaganipalli, C.K. Palli, Dharmavaram, Bathalapalli, Tadimarri, Mudigubba, Anantapur, Kudair, Pamidi and Peddavaduguru in the centre which are mainly made-up of arid treeless expense of poor Red soils.
3. High level land of Penukonda, Roddam, Somandepalli, Hindupur and Lepakshi, Chilamathur, Madakasira, Rolla, Gudibanda and Agali, which connects with Mysore Plateau at higher elevation of the rest of the district. This part has average sandy red soils of normal productivity.

Hills

The Forests in the district are thin and scanty. The Mutchkota hills about 35 km in length run from north of Gooty town up to extreme Southern corner of Tadipatri and Yadiki mandals. Another line of Hilts starts from west of Gooty mandal and run 80 km called by name Nagasamudram hills. The Mallappakonda range begins at Dharmavaram and runs into Karnataka State.

The Penukonda range which starts in the South of Dharmavaram through Penukonda and Hindupur proceeds to Karnataka State. In Madakasira, the hill divides Rolla and Agali mandals into Southern and Northern portions.

There are numerous isolated peaks and rocky clusters, which are devoid of any vegetation. The height of some of these hill-ranges is given below:

Mallappakonda 4 Miles to North of Bukkapatnam	3002 feet
Penukonda	3091 feet
Kundurpi Durgam	2996 feet
Madakasira Hills	2936 feet

Rivers

The principal rivers flowing through the district are Pennar, Chitravati and Hagari and their tributaries. The important river in the district is Pennar. It arises in the Chennakesava hills (or) Chennarayabetta northwest of Nandidurga in Kolar district of

Karnataka State. It is also called Uttar Pinakini and enters this district in the south of Hindupur (Mandal). It is joined by the Kumudvati, its first significant tributary near Hindupur town and by the Jayamangali near Utukuru past Basavanapalli in Parigi mandal. From about Chinnamareddipalli it flows north and assumes a spontaneous course. It continues its northward flow through Roddam (mandal) and almost grazes the interstate border near about Upper Pennar reservoir is constructed over it near Perur (village). Then it passes through mandals Kambadur, Kalyandurg, Beluguppa. At Venkatampalli an anicut is built across the river. It turns east after continuing northwards in the hilly country between Udripikonda and Pennahobilam villages. A few kilometres downstream the mid Pennar dam is built beyond Ramapuram (village) in this stretch the river forms a boundary between the mandals of Beluguppa, Uravakonda, Vajrakarur, Pamidi, Peddavadugur on one side and Atmakur, Kuderu, Garladinne, Singanamala on the other side. It touches its Northern most point (10^0 58') in its entire course at Tadipatri almost parallel to Gandikota range. It then contacts the neighbouring Cuddapah district about 2.4 km to the southwest of Kodur village. Pamidi and Tadipatri are the only important towns on its banks. It is an important source of irrigation in the district. The majority of the channels taken up from it are in Hindupur, Parigi, Lepakshi, Roddam, where one can see some green vegetation along its banks.

Chitravati is the second important river in the district. It rises in Harihareswara hills in Kolar district (Karnataka State) and enters the Chilamathur (mandal) of this district where the elevation of the construction is 846 m. It collects its first significant tributary the Kushavati and penetrates through mandals Gorantla, Puttaparti, Bukkapatnam, Kothacheruvu, Chennakothapalli and Dharmavaram. The river is dammed lower down to form the large tanks of Bukkapatnam and Dharmavaram. It broadens out considerably after gathering the vanka from the Gutturu tank in Dharmavaram (mandal) past Thippepalli it runs east and traverses a little distance before moving northeast near Dhampetla. It provides Dharmavaram Bathalapalle, Tadimarri mandals with good irrigation facilities through its spring channels. It continues in northeast direction only and the Maddileru tributary. And then it approaches the northern fringes of Palakondas and contacts for the first time, Cuddapah district. It once again enters Anantapur district, Tadimarri (mandal) and traverse through Yellanur and touches Cuddapah district.

The Hagari (or) Vedavati river is also an important one in the district has its origin in Karnataka state. It touches this district and traverse through by forming boundary of Gummagatta and Rayadurg on one side and Brahmasamudram and Beluguppa on the other side. Then passes through mandals Kanekal and Bommanahal and enters the Karnataka state. The Chinnahagari between the village of Benakanapalli and Singanapalli joins it. The Bhairavanithippa project (B.T. Project) constructed on this riser affords irrigation facilities to the above said mandals. The river not only feeds the big tanks of Kothapalli, Kanekal and Bhupasamudram and also ensures supplies to the riser channels dug from its bank.

Papagni is another riser, which flows through Tanakal Mandal and N.P. Kunta Mandal. It arises in the Nandi hills of Karnataka state and joins the Pennar River near Adinimayapalli in Cuddapah district. A reservoir is constructed to irrigate over 900 acres.

Apart from these, streams like Kushavati in Chilamathur (Mandal), Swarnamuki in Agali (Mandal), Maddileru in Nallamada, Kadiri, Mudigubba mandals, Pandameru in Kanaganapalli, Rapthadu, Anantapur, Bukkarayasamudram and Singanamala mandals, Papagni in Tanakal mandal are important water supply sources to various large and medium irrigation tanks in the district.

The excavation of river channels often involves considerable labour and expense. The local practices connected with it reflect the salient features of the old Kudimaramathu Act. The small surface flow in the rivers and streams is diverted by means of temporary sand dams during the rainy season and when the rivers become dry channels are excavated with large wooden spades locally called Goraluf drawn by bullocks. They are annually cleared of their silt and sand by the united efforts of the local landholders.

Spring channels which are the interesting features of the irrigation pattern of the district mostly abound in erstwhile Anantapur, Kalyandurg, Dharmavaram, Kadiri taluks and Yadiki Mandal. They are taken from hill-streams, valleys, springs in tank beds and in some cases in ordinary dry lands. Although more dependable than some of the tanks, they irrigate only small areas.

But what is striking is the contribution made towards irrigation by wells in almost all the Mandals. They sustain in particular, the erstwhile taluks of Kadiri, Dharmavaram, Kalyandurg and Anantapur. With the extension of power to agriculture and the popularisation of well-subsidy scheme by Government their number has steadily grown in recent years.

High Level Canal

In 1956, an agreement was arrived between the two states for the purpose of sharing the water of High-level canal in the ratio of 35 to Karnataka state and 65 to Andhra Pradesh. In the first stage of the scheme, estimated to cost Rs. 13 crores involved the excavation of main canal from the Tungabhadra dam up to Uravakonda cut, about 116 miles long out of its first 111 km lies in the Karnataka state. The canal below the 11 km and up to 187 km is expected to irrigate 14,164 ha (35000 acres) in Andhra Pradesh traversing a flat country crosses the Chinna Hagari and Pedda Hagari by means of aqueducts. Below its 187 km, it per acres the Hagari-Penneru watershed through the Uravakonda cut about 8.8 kms the canal is led into the Pennahobilam vanka which falls into the Penneru. The waters are picked up by means of regulator across the Pennar near Penakacherla, which is called the mid Pennar regulator in view of the Upper Pennar project near Perur. The regulator is located upstream of Konamanavanipalli(v) and has two canals, one on the North irrigating 5465 ha (13.500 acres) and another on the south bringing in 28,577 ha (70,615 acres). The north canal is about 54.7 km long almost the whole of its running in Pamidi and Gooty mandals. The south canal runs 60 miles in Garladinne, Anantapur, Bukkarayasamudram, Narpala, Peddapappuru, Putluru and Yellanur mandals of this district, directly serving an ayacut of 16,430 (40,600 acres) and terminates in the Narepalli vanka. The south canal at second stage is designed to carry sufficient water for serving and additional area of 22,258 (5500 acres) in Cuddapah district. The distribution of the various sources in the district and the areas irrigated by them are given below in table 1.

Table 1 Major and Medium Irrigation

Name of the Project	*Nos.*	*Mandals*	*Villages covered (in Nos)*	*% to total area irrigated*
1	2	3	4	
Tungabhadra Project				
High Level Canal Stage - I (1968)		D. Hirehal	3	31.49%
Estimated Cost (Rs. in lakhs) 1174.00		Bommanah al	14	
Expenditure (Rs. in lakhs)1376.00		Vidapanaka l	12	
Ayacut localised (lakh acres) 1.13		Anantapur	3	
Ayacut developed (lakh acres) 0.99		Garladinne	15	
		Narpala	6	
		Yellanur	5	
		Kanekal	7	
		Uravakonda	3	
		Vajrakarur	1	
		Singanamal a	9	
		Bukkaraya samud ram	10	
		Tadimarri	9	
		Putlur	10	
		Peddapapp ur	1	
	Total		**108**	

2. Thungabhadra Project				
High Level Canal Stage - I (1968)				
Estimated cost (Rs. in lakhs) 635.00				
Expenditure (Rs. in lakhs) 618.00		Gooty	12	
Ayacut localised (lakh acres) 0.32		Pamidi	18	
Ayacut developed (lakh acres) 0.20		Guntakal	6	
		Peddavaduguru	8	
	Total			

Medium of Irrigation

1. Bhairavanithippa Project				
Date of completion 1961				
Expenditure (in lakhs) 143.74		Gummagatta	8	
Extent localised (in acres) 12.000		Bramhasamudram	6	
Extent developed (in acres)9.390				
	Total		**2**	

Upper Pennar Project				
Date of completion 1958		Kambadur	3	
Estimated cost (Rs. in lakhs) 160.66		Kalyandurg	3	
Expenditure (in lakhs)		Rayadurg	3	
Extent localised (in acres)		Rapthadu	4	
Extent developed (in acres) 8372.00				
	Total		**1**	

Chennaravaswamv Gudi Project (C.G.P)				
Date of completion 1960	3	Nallacheruvu	4	
Estimated cost (Rs. in lakhs)				
Expenditure (in lakhs)				
Extent localised (in acres)				
Extent developed (in acres)				

(a)Above 200 (New) (Acres) 4429	10			
(b)100 to 199 (acres) 44.303	282			
(c)Old (acres) 63.752	129			

Minor Irrigation (Panchayat Raj)				
(a)Tanks (in acres)	1071			9.74
(b) Others sources such as Nallas, Springs, River channels, etc. (in acres) 52.404.20	1658			2.98
(a) Wells and filter points 674				3.28
(b) Other wells 60811				52.51
(c) Electric Motor Pumpsets 57.388				100

Anantapur is the southern-most district of the Rayalaseema region of Andhra Pradesh. While agriculture remains the most important economic activity of the district, it is characterised by high levels of instability and uncertainty. Being located in the rain-shadow region of Andhra Pradesh, the district is drought-prone. By now, it is well established that while a generalised rural crisis is prevalent across the country, a disproportionate burden has fallen on its drier tracts. In 2006, Anantapur was one of the thirty-one districts identified by the Government of India as being prone to agriculture-related suicides. A few committees have been set up by the Government of Andhra Pradesh to study the agricultural crisis as well as the problem of farmers' suicides in the state. While these committees have looked at a whole gamut of issues underlying the agricultural crisis of Andhra Pradesh, our study of Anantapur district focuses specially on the technological dimension of the crisis. Our study addresses such factors that have a bearing on agricultural production and productivity namely, the physical environment, the extent and nature of agricultural inputs used the nature of crop protection practices followed; the overall management practices adopted in cultivation including irrigation practices and aspects of technology delivery.

Anantapur district is bounded by Kurnool district in the north, Cuddapah district in the northeast, Chitoor district in the southeast, and Karnataka State on the west (Map 1). The district has a total geographical area of 19.13 lakh hectare. For administrative purposes, the district is divided into three revenue divisions, namely, Anantapur, Dharmavaram, and Penukonda, there are sixty-three revenue mandals. As per 2001 census, the district has 10 towns and 964 revenue villages and a total population of 3.64 million. Almost 75 percent of the population in the district lives in rural areas. Agriculture remains the predominant activity in the villages, with 80 percent of total workers engaged in agriculture, either as cultivators or

Map 1 Anantapur District

agricultural labourers. In urban areas, about 11 percent of the workforce is engaged in agriculture. Mining is also an important activity in Anantapur district as it is endowed with rich deposits of iron ore and lime stone, as well as other minerals. There are more than fifty small-scale industrial units in the district, of which nearly one-half are to do with granite. There are also couple of cement industries and steel industries in the district.

The impact of drought on area, production and productivity in Anantapur district revealed that among the respondent sampled households, the acreage of cereals (jowar, bajra, korra and arike) declined by 16 percent and production by 31 percent during the drought year in 1982-83 compared to the preceding year. Production of groundnut sharply declined by 27 percent and productivity by 40 percent respectively (Naidu 1985: 790).

Rayalaseema is one of the three major geographical regions of Andhra Pradesh. The other two regions are Coastal Andhra and Telangana. The Rayalaseema region of Andhra Pradesh comprises of four southern districts of Kurnool, Anantapur, Cuddapah and Chittoor. It has an area of about 73,495 sq km, and forms 24.46 percent of total area of Andhra Pradesh. The population of Rayalaseema region is 116.86 lakhs according to 1991 census. It accounts for 17.6 percent of total population of Andhra Pradesh (Census of India 1991: 85). The region lies in between the north latitudes 12^0 30' and 16^0 20' and east longitudes 76^0 13' and 80^0.

Drought-Prone Nature

Monsoon failures have been recurring phenomena in many parts of India. Hardly a year passes in which some part or other of the country does not, in some degree, suffer from the calamity of drought. The most disastrous droughts come at irregular intervals. The core areas of drought comprise about 16 percent of the total geographical area of the country and account for 11 percent of its population[2] (Ministry of Irrigation and Power 1972: 157). Starvation, migration and such problems have been a part of the lives of people in these regions.

The criteria for declaring "drought" have been variously debated. However, two criteria adopted by Government of India - rainfall deficit and available irrigation facility seem to be reasonable. It is observed that, "Areas where the frequency or probability of failure of annual rainfall by more than 25 percent from the normal was found to be 20 percent or more for the observed years, were considered as drought-prone. Areas where the frequency exceeded 40 percent were considered as chronically drought-prone"[3] (cit. in Nadkarni 1985: 24).

However, the emphasis on total rainfall may not be correct. The even distribution of rainfall and the number of rainy days are important. Even if the total rainfall received is normal, lack of its proper distribution vis-a-vis plant requirements will lead to failure of crops. In regions like Gujarat, Rajasthan or Andhra Pradesh, where rainfall is highly deviant from year to year, "drought" can be shown in any district in any year. In fact, in acutely drought affected regions like *Rayalaseema*, studies (on the basis of rainfall data since 1945) have indicated that no meteorological drought had occurred.[4]

The *Rayalaseema* region is historically known as "stocking ground of famines". Anantapur district located in Rayalaseema is the driest among all the districts of Andhra Pradesh and is drought-prone. The Irrigation committee (1972) has identified the entire district as drought-prone. It is estimated that drought visits Anantapur district every alternate year. A single dry crop is raised in most parts of the district under rainfed conditions. Anantapur district is one of the 6 districts in India known for their low average rainfall, frequency and severity of droughts and low proportion of the irrigated land to total cultivated land. Consequently, the region is rated low for its level of socio-economic development. The World Bank has also bestowed its attention in this region because of its drought-prone nature and its severity.

To have a better insight about drought proneness of Anantapur district, a few details about the district and also other Drought Prone Districts (DPD's) and Non-Drought Prone Districts of Andhra Pradesh are presented here.

In more than one-half of the years studied, the actual rainfall is below the annual mean rainfall of 56-8.5 mm. That is, fifty-one out of ninety-four years have experienced below mean rainfall in

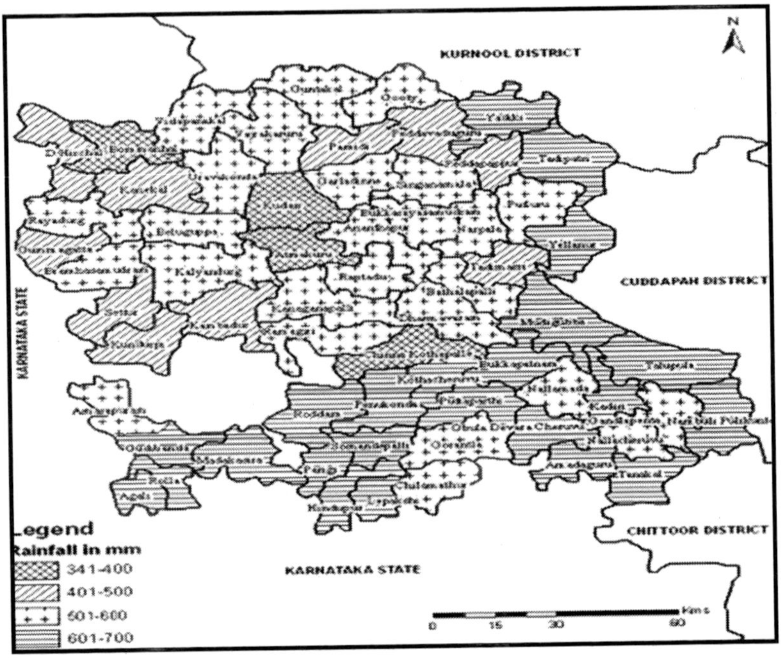

Map 2 Mandals Classified by Normal Rainfall - Anantapur District

Anantapur district (fig. 1). Further, on average once in every five years, the district experiences drought conditions, eighteen out of ninety-four years are classified as drought years, as the annual actual rainfall in these years has been 75 percent below the annual mean rainfall of 568.5 mm. Fourteen out of these eighteen years are moderate-drought years while four may be classified as severe-drought years.

An analysis of the contribution of seasonal rainfall to total annual rainfall of the district, from 1911 to 2004, shows that the southwest monsoon (June to September).

It contributes 58 percent, northeast monsoon (October to December) 28 percent, hot weather period (March to May) about 13 percent, and cold weather period (January and February) 1 percent. Normal rainfall shows a marginally declining trend across all the seasons in the district over a ninety-year period, located in the interior Deccan plateau, the district has warm and dry climate, with a very low annual rainfall of 544 ram against 891 mm in the state. The

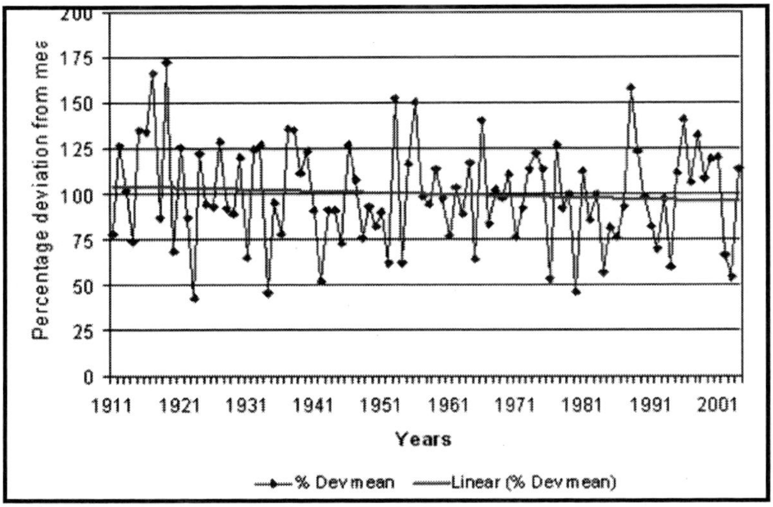

Fig. 1 Percentage Deviation from Mean Annual Rainfall - Anantapur District

According to the India Meteorological Department, meteorological drought occurs when the actual annual rainfall is less than or equal to 75 per cent of the mean annual rainfall. Severe drought occurs when the actual annual rainfall is below 50 per cent of the mean annual rainfall; moderate drought occurs when the actual annual rainfall is 50 to 75% of the mean annual rainfall.

variation in normal rainfall across taluks is not very wide, ranging from 499 mm in Kalyandurg to 617 mm in Kadiri. The district has the lowest rainfall in Andhra Pradesh and even at all-India level it is second lowest.

In addition to the marginally declining trend of the quantum of rainfall, variability in rainfall is also an important issue affecting the agricultural prospects of Anantapur district. Variability in rainfall is noticed in all seasons, except the southwest monsoon, as testified by the values of the coefficient of variation calculated over 1911 to 2004. That is, on average, about 42 percent of annual rainfall is subject to a high degree of variability. However, rainfall is assured during the southwest monsoon, though with a low mean of 329.9 mm. Yet, the effective rainfall which indicates the part of the rainfall that is available as soil moisture in the root zone to meet the crop evapo-transpiration requirement) is computed to be 198 mm for southwest monsoon over the period 1911 to 2004. The effective rainfall in the district is at least 50 percent below the amount required to reap potential groundnut yield. Further, comparing mean rainfall and potential evapo-transpiration (PET)[8] indicates that during the main crop-growing period of south west monsoon, quantum of rainfall is lower than the levels of potential evapo-transpiration in the district. The rainfall deficit with respect to PET was 70 percent in June, 64 percent in July, 52 percent in August and 2 percent in September. The soil-moisture stress condition under different stages of crop growth would result in inadequate plant population, higher percentage of flower drop, poor seed setting, etc. and thereby have implications for crop yields.

Table 2 Average Annual Rainfall in Anantapur District, Drought Prone and Non-Drought prone Districts of Andhra Pradesh

District	Rain fall (in mm) 1965-66	1976-77	1986-87	Normal
Anantapur	368.50	460.00	439.03	544.03
Drought Prone Districts	475.03	759.00	521.00	693.03
Non-Drought Prone Dist.	766.15	1172.00	1040.00	1033.03

Source: Census Reports of respective years, Government of India

As it can be seen from the table 2 that, 1976-77 being a non-drought year in general, Anantapur received only 460 MRAI of rainfall as against 759 in Drought Prone districts and 1172 in Non-Drought Prone Districts, in both the latter cases it was above the normal rainfall. This indicates the precarious rainfall situation that is prevalent in Anantapur district.

Table 3 Normal Rainfall (in mm) Across Seasons - Anantapur District

Period	South-West monsoon	North-East monsoon	Cold weather Period	Hot weather Period	Annual rainfall
1911-40	350.27	166.57	13.58	72.93	603.35
1941-71	315.97	148.88	2.06	84.35	551.26
1971-2000	335.58	157.51	2.06	64.30	559.46

Source: Data collected from IMD, Pune

Soils

The district is covered mostly (80%) by red soil. However, three distinct natural divisions can be identified in the district. The northern part of the district extending from Uravakonda to Tadipatri has predominantly black cotton soils. Though the soil is fertile, its sloppy nature leads to soil erosion and low water retention. The central part has infertile red soil with an admixture of black soils in parts. The southern part of the district has superior red soil. This part is less sloppy and hence has more cultivation and vegetation.

Soils differ greatly in their morphological, physical, chemical and mineralogical properties. These differences affect the response of the soils to management inputs. For appraising the productivity of the soils and for the preparation of perspective plan for land use and soil conservation for sustainable development one needs to have thorough knowledge of the kind and distribution of different soils of the area. The classification of soils of Anantapur district according to Soil Taxonomy (USDA: 1975) places them in the orders of Entisols, Inceptisols, Vertisols and Alfisols. These further divided into sub-

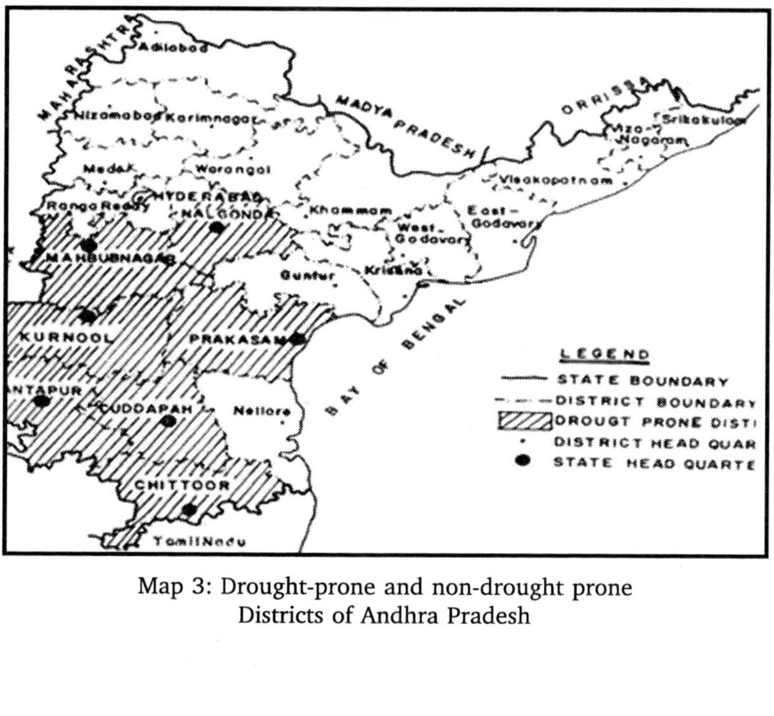

Map 3: Drought-prone and non-drought prone Districts of Andhra Pradesh

order great groups and sub-groups in each of the orders, which have a number of soil series.

Table 4 Taxonomic Classification of Soils of Anantapur District

S.No.	Order	Sub-order	Great Group	Sub-Group	Series
1	Entisols	Orthents	Ustorthents	Lithic Ustorthents Typic Ustorthents Aquic Ustorthents	Peddamallepalle Chikkahagari Garudapuram Sivaram
		Fluvents	Ustifluvents	Typic Ustifluvents	Utakallu Manesamudram Anantapur Hagari Pennar Tamballapalle Kunuturu Kushavati Jayamangale Uravakonda
				Molli Ustifluvents	Gollapalli
2	Inceptisols	Ochrepts	Ustochrepts	Lithic Ustochrepts	Velidandla Kadavakallu Kurugunta Aveulena Virupapuram Kondapuram
				Lithic Vertic Ustochrepts	Itigi
				Vertic Ustochrepts	Chitravati
				Fluventic Ustochrepts	Rachepalle Batrepalle Tadapatri Peruru Vajrakarur
3	Vertisols	Listens	Chrobmusterts	Typic Chrbmusterts	Bogalakatta Hugalur
			Pellusterts	Typic Pellusterts	Mutssukota, Poliki Roddam
4	Alfisols	Ustalfs	Haplustalfs	Typic Haplustalfs	Cherlopalle Kalyam Mancheru Satralapalle Kottacheruvu

contd....

5			Rhodustalfs	Udic Rhodustalfs	Devalapur Gooty Talak Kanapuram Kadiri Madigapalle Ramapuram Vayalpad

The Soils in Anantapur district are predominantly red except in mandals Kaneka, Bommanahal, Vidapanakal, Uravakonda, Vajrakarur, Guntakal, Gooty, Pamidi, Peddavadugur, Tadipatri, Yellanur, Yadiki, Peddapappur and Putlur. These mandals occur with red and black soils almost in equal proportion. Thus, 76% red and 24% black soils.

Soil is mostly from disintegrated gneisses though occasional quart zone formations are seen here and there. On the whole the soil in the division is more or less red-sandy ferruginous loam and shallow in depth and mostly the forest areas are heavily grazed and subject to annual fires. The surface is devoid of any humus except in a few sheltered valleys. In some rare plain portions and valleys, deep sandy to clayey loam is met with. The depth of soil in such localities gradually decreases as it approaches the hill slopes, where it tends to be loose and boulder. It is often gravelly and occasionally mixed with boulders of varying sizes and varies from red sandy ferruginous to clayey loam. The topsoil in some of the plainer portions of the panchayat reserve forests is saline. Black cotton soil occurs in Gooty range especially in North Western and Eastern portions of it. Sandy soil occurs in Yellanur reserve forests of Gooty range.[5]

Mineral Wealth

Anantapur is one of the two gold bearing districts in Andhra Pradesh. Gold occurs in Ramagiri mandal of Anantapur district, associated with white to bluish quartz veins traversing in the Dharwarian schist. The ore shoots occur in a linear north-south belt extending over a length of 11 km with a width of 1.5 km.

Diamonds occur in Kimberlite pipes of Vajrakarur Lattavaram belt. Seven such kimberlite pipes have been recognised (Anon. 1993).

These pipes are intrusive into the granites and gneisses hosting diamonds.

In addition to gold and diamonds other minerals that occur in Anantapur district are barite, steatite, iron and granite. Snow white and off-white barite veins are observed as fault fillings in the Vempalli dolomites near Mutsukota. The best variety (lava grade) of steatite is also being mined in this area. Small pockets of iron ore (hematite) occur in Rayadurg area. Black, pink and multi-coloured granites are available which are used in the cutting and polishing industry.

Land-use Pattern

Due to the persistent drought, the area under current fallows in drought prone districts is always higher than that of Non drought prone districts and this is all the more true in respect of Anantapur district, which is hard hit by the droughts. The share of cultivable waste and other fallows also show a similar trend during 1965-66 to 1986-87 (see table 2).

Table 4 Structure of Land Holdings in Anantapur District, 2001

Size Class of Operational Holdings	*Number of Holdings*	*Area in Ha.*	*Area under Current Fallows in Ha.*	*Current Fallows as Percentage of total area*	*Average Size of Holdings (in Ha.)*
Marginal	206,326	113,902	12,287	10.79	0.55
	34.46	9.51			
Small	185,705	272,953	18,109	6.63	1.47
	31.02	22.78			
Semi-Medium	146,238	373,874	23,497	6.28	2.56
	24.43	31.21			
Medium	51,434	299,541	23,870	7.97	5.82
	8.59	25.00			
Large	8,969	137,726	15,443	11.21	15.36
	1.50	11.50			
All classes	598,672 100.00	1,197,996 100.00	93,205	7.78	2.00

Note: *1. Marginal holdings are below 1 Hectare; Small holdings between 1 to 2 hectare; Semi Medium between 2 and 4 hectare; Medium are between 4 and 10 hectare and Large are above 10 hectare.2.Figures in bold are percentages with respect to the total.* ***Source:*** *http://agcensus.nic.in/ cited on 06-09-08*

In the year 2001, as Table 4 illustrates, 7.78 percent of total area under land holdings is classified under current fallows. Farmers in both ends of the spectrum, marginal and large, record a high percentage of land as fallow, namely, 11 percent, while farmers in other size classes also leave a considerable extent, 6 percent to 8 percent, as follows. While the exact reasons for withdrawal of land from cultivation would vary from one farmer to another, considering the high magnitude of degraded land that is prevalent in the district, it is reasonable to assume that treatment of degraded land would help reduce the extent of fallows across all size classes of farmers. The relationship that prevails between land degradation and fallows thus has an important policy implication. By addressing the issue of land degradation, the fallow lands, at least in part, can be brought back to cultivation.

Analysing the land-use classification in our survey villages, it is noted that culturable land that is underutilised, current fallows, other fallows and culturable waste land is more or less of the same order in our survey villages as in the district as a whole in 2005-06 (Table 5). In the whole district as well as in Thirumaldeverpally village, about 15 percent of geographical area remains as fallow and culturable waste while in Thopudurthi, the corresponding percentage is about 12 percent. Thirumaldeverpally village also reports a high percentage of land as high as 20 percent, that is left barren. In Thirumaldeverpally village, the changing irrigation scenario in the village partly explains the high extent of fallows and barren land. Some portion of the culturable land in the village used to receive irrigation from a stream as well as from a rain-fed tank. Over the years, while the stream has more or less dried up, irrigation from the rain-fed tank is quite uncertain. The portion of culturable lands in the village that used to receive irrigation from the stream and tank are predominantly covered by black soil which is not particularly suitable for groundnut cultivation. Therefore, with the decline in the water flow in the stream and tank, farmers are forced to dig tube wells to continue to cultivate this land. If a farmer is unable to successfully dig a bore well, then he is left with no choice but to withdraw this land from cultivation either on a permanent basis or until such time when there is water available for irrigation.

Table 5 Temperature

Month	Mean daily maximum tempera-ture °C	Mean daily minimum tempera-ture °C	Relative Humidity 0830%	1730%
January	30.0	17.3	68	39
February	33.3	18.6	55	8
March	36.9	21.6	48	21
April	38.5	25.6	54	24
May	38.1	25.8	62	1
June	34.9	24.8	68	45
July	32.7	23.8	74	56
August	32.6	23.6	74	54
September	32.2	23.0	75	54
October	31.4	22.3	76	56
November	29.9	19.2	70	49
December	28.7	16.8	72	46
Annual	33.3	21.9	66	42

Table 6 Agricultural Statistics of Anantapur District 1988-89
Total agricultural land 13,63,443 hectares

S.No.	Land use /Land cover category	Area in hectares	Percentage to total geographical area
1	Kharif irrigated	2,54,611	18.67
2	Kharif unirrigated	9,70,848	71.20
3	Rabi irrigated	2,27,147	16.66
4	Rabi unirrigated	1,03,446	7.59
5	Double cropped area	2,22,468	16.32
		13,33,584	97.81
6	Net area sown	—	—
7	Gross cropped area	15,56,052	—
8	Fallow land	18,435	1.35
9	Plantation	11,424	0.84

Forests

It is an area (within the notified forest boundary) bearing an association predominantly of trees and other vegetation types capable of producing timber and other forest produce.

Deciduous Forest (Dense)

It is described as a forest, which predominantly comprises of deciduous species and where the trees shed their leaves once in a year. The vegetation density (crown cover) is 40% or above, it is called dense or closed forest. This type occurs in a localised manner in small pockets in the valleys, hill slopes and near the perennial water source areas where with plenty of humus and moisture in the soils is present. Deciduous (dense) Forest is identified in Bukkapatnam mandal, Raptadu mandal near Mudigubba and Kadiri mandals.

Deciduous Forest (Open)

It is described as a forest, which predominantly comprises of deciduous species and where the trees shed their leaves once in a year. The density (crown cover) is 20-40%. It is called open forest. There is no open forest in the district.

Degraded/Under-utilised Forest

It is described as a forest where the vegetative (crown) density is less than 20% of the canopy cover. It is the result of both boitic and aboitic influences. Most of the forest in Anantapur district is identified as degraded under-utilised forest. This type of forest has been identified near Rayadurg, Guntakal and Nallacheruvu mandals, etc.

Forest Plantations

It is described as an area of trees of species of forestry importance and raised on notified forestlands. It includes eucalyptus, casuarinas, etc. Eucalyptus and casuarinas plantations have been identified in this district.

Waste Lands

It is described as degraded land which can be brought under vegetative cover with reasonable effort and which is currently under utilised and land which is deteriorating due to lack of appropriate water and soil management or on account of natural causes. Wastelands can result from inherent imposed constraints such as by location, environment, chemical and physical properties of the soil or financial or management constraints. The wastelands identified in the district are described below:

Salt-affected Land

The salt-affected land is generally characterized as the land that has adverse effects on the growth of most of the plants due to the action or presence of excess soluble or high exchangeable sodium. Alkaline land has an exchangeable sodium percentage (ESP) of about 15, which is generally considered as the limit between normal and alkali soils. The predominant salts are carbonates and bicarbonates of sodium. Coastal saline soils may be with or without ingress or inundation by seawater. These lands have been identified in Anantapur, Peddavaduguru, Gooty, Dharmavaram and Kundurpi Mandals.

The gullies are formed as a result of localised surface run-off affecting the friable unconsolidated material in the formation of perceptible channels resulting in un-consolidating terrain. The gullies are the first stage of excessive land dissection followed by their networking which leads to the development of ravenous land. The word ravine is usually associated not with an isolated gully but a network of deep gullies formed generally in thick alluvium and entering a nearby river flowing much lower than the surrounding high grounds. The ravines are extensive system of gullies developed along river courses.

Land with or without Scrub

They occupy (relatively) higher topography like uplands or high grounds with or without scrub. These lands are generally prone to degradation or erosion. These exclude hilly and mountainous terrain.

These lands are identified in Yadiki, Tadpatri, Narpala, Vajrakarur, Kudair, Dharmavaram, Ramagiri, Penukonda, Puttaparthi, Gorantla, Chilamathuru, Mudigubba and Gandlapenta Mandals.

Sandy Area

These are the areas, which have stabilized accumulation of sand in situ or transported in coastal riverine or inland (desert) areas. They occur either in the form of sand dunes, beaches, channel (river/ stream) islands, etc.

Mining and Industrial Wastelands

Lands where large-scale mining operations bring the degradation of land and resultant mine dumps are classified under this category. Industrial wastelands, which subject to degradation caused by large scale industrial effluent discharges.

Barren Rocky/Stop Waste/Sheet Rock Area

It is defined as the rock exposures of varying lithology often barren and devoid of soil cover and vegetation. They occur amidst hill forests as openings or scattered as isolated exposures or loose fragments of boulders or as sheet rocks on plateau and plains. These areas have been identified and mapped in Ramagiri, Tadpatri, Yellanur, Tadimarri, C.K. Palli Mandals.

Water Bodies

It is an area of impounded water areal in extent and often with a resulted flow of water. It includes man-made reservoirs/lakes/ tanks/canals besides natural lakes, rivers streams and creeks.

River/Stream

It is a natural course of flowing water on the land along definite channels. It includes from a small stream to a big river and its branches. It may be perennial or ephemeral (all the rivers/streams have been identified and mapped).

Reservoirs/Lakes/Tanks/Canals

It is natural or man-made enclosed water body with a regulated flow of water. Reservoirs are larger than tanks/lakes and are used for generating electricity, irrigation and flood control. Tanks are smaller in a real extent with limited use than the former. Canals are inland waterways used for irrigation and sometimes for navigation.

The percentage share of the forests in the total geographical area is one of the indicators of any district. Forest coverage in Anantapur is lowest of all the districts of Andhra Pradesh. Area under forest cover is only around 10 percent in Anantapur district, whereas in other Drought Prone Districts it is 17 to 20 percent.

Map 4 Mandal-wise Ground Water Resource, Anantapur District
(Source: Department of Groundwater, Anantapur)

Analysing district-level data on irrigation, it is clear that the percentage of area irrigated was about 10 percent in the early 1960s and reached a peak of 16 percent to 17 percent during late 1980s. Then again there was a decline, and by the mid-2000s the percentage of area irrigated was only 11 percent to 12 percent. Gross irrigated area as a percentage of gross cropped area shows a similar trend too. While an analysis of the extent of area under irrigation, over the period 1960-61 to 2006-07, shows an overall increasing trend, there has been a decline in the extent of area under irrigation since late 1980.

British Period History Famines

The history of famines in a district where scarcity is almost the rule is bound to be rather tedious but some of them which were either prolonged or intense merit description. Although elaborate evidence wants in regard to the havoc that some of the earlier famines brought upon the district before its formation in 1882. The review written in 1886 by Nicholson, the Acting Collector of Bellary[6] serves to show how frequent and unfortunate the distresses were between 1803 and 1885 and the following paragraphs discusses the major calamities.

Rainfall

The rainfall of Anantapur is tighter than that of any other district in the Presidency except Bellary and the difference between these two is only a matter of a few cents the annual fall in both of them averaging between 22 and 23 inches. Anantapur lies midway between the two monsoons but receives the full benefit of neither. It is too close to the Western Ghats to profit greatly by the southwest monsoon and the northeast current has left behind most of its moisture before reaching it.

There are three fairly distinct zones of rainfall in the district: The three driest taluks are Anantapur, Dharmavaram and Kalyandurg in the centre, where as in the adjoining Rayadurg and Bellary taluks the total received averages less than 21 inches, Gooty and Tadipatri receive about 23 inches and in the three southern taluks, where the rain is less scanty than elsewhere, between 23 and 24 inches is registered, for the district as a whole the heaviest fall on record is that in 1874 the year before the beginning of the great famine of 1876-78, when 38.6 inches was received. The lightest was that in 1876, during that famine, when the district average was only 6.9 inches. Since 1870 the fall has exceeded 25 inches in thirteen years and been over 30 inches in four. These four were 1874, 1878 (38 inches), 1889 (33 inches) and 1903 (32 inches).

Table 7 Average Rainfall for the years 1870-1903

Station	*January to March*	*April to May*	*June to September*	*October to December*	*The whole Year*
Gooty	0.11	2.15	15.04	6.01	23.31
Cra' akonda	0.27	2.72	13.15	6.25	22.39
Tadpatri	0.20	1.60	14.58	6.62	23.00
Yadiki(Rayalacheruvu)	0.20	1.98	13.83	5.70	21.71
Anantapur	0.17	2.23	12.35	6.03	20.78
Dharmavaram	0.29	1.93	11.71	6.77	20.70
Kalyandurg	0.28	3.02	11.15	6.03	20.48
Penukonda	0.30	3.36	12.49	8.01	24.16
Bukkapatnam	0.34	1.94	14.79	7.65	24.72
Hindupur	0.33	3.33	12.22	7.36	23.24
Madakasira	0.28	4.07	11.79	8.11	24.25
District average	0.25	2.69	13.05	6.93	22.92

Possessing therefore, an exceedingly light and withal uncertain rainfall, a district in which three-fourths of the people are dependent upon pastoral and agricultural pursuits, where the soil is most of it poor and where the irrigation works are almost all of them dependent upon local rainfall. Anantapur is a necessary corollary, more than usually liable to disastrous seasons. The official records of the bad years, which occurred in it before it was severed from Bellary in 1882 usually, however, deal with the old Bellary district as a whole and it is not always possible to give separate detailed particulars for the taluks, which now make up Anantapur.

Of the famines and scarcities, which overtook it before its cession to the company, there is no exact record. Indigenous historians in those days concerned themselves more with courts and kings than with calamities of the common people.

One of Munro's reports makes passing mentions of a scarcity in 1756 and shows that the famine of 1791-92 was so severe in the Northern Circars (and which is memorable as being the first occasion. On which an Indian Government opened relief works) also seriously affected the Deccan district. Its intensity was enhanced by the capacity of the native administration. Had the officers of Government, says Munro, lowered the assessment or even let it remain as before. The effects of the famine would probably only have been felt while it lasted, but as they raised it near 50 percent wherever there was a crop. This addition to the high price necessarily occasioned by the scarcity rendered grain so dear that very little could be purchased by the lower classes of the inhabitants, and great numbers of them perished in consequence.

Famine of 1802-1804

The southwest monsoon had failed in Rayalaseema during June 1802 resulting in the cessation of agricultural operations. Munro, the Principal Collector, informed the Board of Revenue that in no one village perhaps of any district except Adoni was thus what is called an average produce.[7] Rains failed again during the Northeast monsoon season producing conditions of drought. The drought deepened into famine when the southwest monsoon proved deficient for the second consecutive year in 1804. Scarcity of food caused a steep rise in the prices of grain. Munro sent an alarming report to the Board of Revenue on the state of the prices. He wrote: Rice, which formerly sold here from twenty-five to thirty seers a rupee, is now selling at eight and nine seers.[8]

In disregard of the famine conditions grain was exported to Raichur doab in the Nizam dominion where the price of grain was higher than that in Rayalaseema. The suggestion of Munro to impose an embargo on export of grain was not given effect to. Consequently, the distress of the poorer sections of the society deepened. The poorer ryots were seen migrating to the neighbouring Mysore state.[9] Even though statistics on deaths resulting from this famine are not available, one may safely conjecture that the rate of mortality of people and livestock might have been great.

A year later he wrote again the Kanks nowhere received any water for first crop. In only two out of 37 districts did they receive a full supply for the second? Great numbers of the cattle employed in agriculture and in the transport of grain have perished. All kinds of provisions are from two to three hundred percent above the average rate. The scarcity (Munro declined throughout to describe it as a 'famine' ended with the great storm of October 1804.

Floods

The first serious floods in the district were that of October 1804. Writing to the Board on the 4th November of that year Munro said, In consequence of a torrent of rain between the 12th and 15th of last month, during which all the rivers and nullas rose to a height never before remembered, the greater part of the tanks have been destroyed over every pan of the country from Harpanahalli to Chitvel.[10] The nullas cut from rivers have been buried in sand and in many places so does that it is difficult to discover their former channels and many villages have been swept away with all the property they contained and in several instances with a considerable number of their inhabitants.

In the next year he sent some more details of the damage done. In the four ceded districts four dams, 752 tanks, 260 "nullas" and 855 wells had been either destroyed or greatly injured and although only such of them had been restored as possessed sufficient ayacut to repay the out lay in four years, as much as lakhs of rupees had been spent in repairing the damage they had sustained.

Famine of 1806-1807

Following the favourable northeast monsoon of 1804 and the southwest monsoon, agricultural operations commenced and much of land was brought under food crops, the crops of the present season. Munro reported are more abundant than those of an ordinary year. However, scarcely a year had passed when in 1806 there was widespread failure of rains in *Rayalaseema* and elsewhere in the Madras Presidency. The resulting drought was so extensive that grain became dear everywhere. It was feared that the prospect of the failure of rains during May-June might further contribute to the rise in the price of grain. Despite the gravity of the situation, grain export was continued which "obliged" Munro to submit to the Board of Revenue the propriety of forbidden, exportation to foreign [neighbouring] states. As the rains failed during the sowing season of 1806, scarcity further deepened in early 1807. Second sort rice was sold in

Rayalaseema between 7 and seers per rupee. The loss of even dry crops such as *cholam*, ragi, etc. was so complete that the farm labourers found it difficult to obtain their usual food since their wages were mostly paid in grains. The production of cotton in the year was computed at one-tenth of the average production of previous years. Munro reported that 10 to 15 percent of the cattle employed in agriculture and about 50 percent not employed in farm activities perished for want of grass.[12] Large crowds of emaciated people flocked into the city of Madras, attracted thither by the existence of a charitable association, and in the expectation of obtaining gratuitous help without any limits. A large number of ryots were forced to sell their cattle to pay the land revenue and join the ranks of farm-labourers.[13] The land revenue collections, however, declined from Rs. 5,374,884 (1805-06) to Rs. 4,489,627 (1806-07).[14] Munro's efforts to grant tuccavi loans to enable the ryots to repair wells and tanks were "disapproved" by the Board of Directors of the Company.[15]

During this famine, it may be noted, there was a debate in the official circles as to the propriety or other-wise of Government interfering in grain trade in order to make food available to the people at a fair price. Munro was of the opinion that encouraging import and prohibiting of food export, by reducing the rents of the lower classes of ryots and by providing employment to the poor on public works could alleviate the conditions of distress. He gave top priority to import of food grains, which he felt would reduce the prices and thus facilitate the poor to purchase food. He argued that in that event, it would be unnecessary to take any steps for the assistance of the poor, because they would easily find employment among the other classes of the inhabitants.[16]

The Government of Madras, in the initial stages of the famine, disfavoured the idea of stopping export and import of grain on Government account. Influenced by Adam Smith's doctrine of freedom of trade, it directed its officers not to interfere in grain trade in any manner whatsoever. It stated that the grain market should be left entirely to him (the grain dealer), i.e. that it he left entirely free.[17] It also desired to encourage the traders to import grain and to employ the poor on public works.[18] However, the incapacity of the grain dealers to meet the requirements of the people did ultimately force the Government to import grain on public account and also guarantee a fixed price for the grain imported by private dealers.[19] The Government's vacillating attitude with regard to purchase and distribution of grain on public account at the beginning of the crisis had nullified whatever benefits might have accrued from the subsequent importation of grain and regulation of its sale price.

This famine demonstrated the vulnerability of the poorer sections of Anantapur region to famine and revealed the inadequacy of Government measures to alleviate their misery.

Relief Measures

The Government relief consisted of remission of land tax and provision of work to the famine-affected people on repair of roads and digging of wells.[20] Private charity was also available to the famine-stricken poor at Madras town during this famine. The Monegar choultry[21] provided food to large number of the poor who came to that place from the neighbouring districts. It was, however, not possible to save the tens of thousands of people froth hunger and disease by a single organisation like the Monegar choultry. Hence the death rate in the city, because of the influx of the famine-affected people from outside had shown a significant increase during 1807.[22]

It may be mentioned that the Government imported 29,010 tonne of rice from the surplus areas of which only 5,943 tonne was consumed. The rest of it had to be disposed off by the Government because the southwest and northeast monsoon rains brought down the prices. Thus the heavy loss suffered by the Government provided an opportunity for the protagonists of Adam Smith and Mill School of political economy to taboo any sort of interference on the part of the Government in food grain trade in all circumstances. The strict adherence to the policy of unrestrained freedom of trade even under conditions of scarcity thus became the hallmark of Government's famine policy in with tragic consequences.

Famine of 1812

The erratic seasons of 1810-11 produced scarcity conditions in the districts of Rayalaseema and Karnataka. The prices of grain rose abnormally causing hardship to the people. The failure of dry crops in most parts of the district for the second year in succession aggravated the situation. The Collector warned the Board that "if the season continues unfavourable the inhabitants of these districts (taluks in the district) will either have to leave them or starve."[23]

The Government's relief policy was confined to exempting grain from customs duties throughout the presidency and affording employment to the poor. The Government favoured giving an advance of money to merchants to import grain from the surplus to the deficit areas, while it forbade, as before, the district collectors from interfering in the grain trade. The Board of Revenue, for instance, disapproved the proposals of the Nellore Collector to coerce the traders to open their stocks of grain. The Board wrote: No dealer or farmer could possibly be supposed to hoard grain for the mere sake of hoarding it. Their intention must always be to sell whenever they could get an adequate price: which price if there were no interference, could not fail to adapt itself to the existing necessity or demand. Although no reliable statistics of deaths from the famine are available, the mortality rate should have been higher during 1811-12 than in the years of normal.

Storm of 1817

Thirteen years later, on the 19th October 1817, the monsoon again set in with great violence writing in the following June. The collector reported, 117 tanks, 58 channels and 312 wells in Bellary and Anantapur had been either breached or seriously injured. In the latter district, the Gooty and Yadiki taluks suffered most. Nearly a lakh of rupees was spent in repairing the damage and two lakhs more were granted in remissions. More over the rain continued to fall for weeks without intermission and consequently hardly any cholam could be sown and what little was put down was drowned by a return of very wet weather in November. So that the whole of the country usually grown with this crop presented one uniform picture of desolation.

Famine of 1823-24

The greater part of the Madras Presidency suffered from severe drought in 1823 creating scarcity conditions and making food dear. The price of grain rose in many places to double its usual rate.[24] The Madras Board of Revenue communicated to the Government on the alarming situation in many parts of the Northern Circars. Karnataka and Rayalaseema districts consequent to the failure of the accustomed rainfall.[25] The Collector of North Arcot, for instance, stated that, the distress of the inhabitants in some parts of the district is very great for want of water for themselves and cattle.[26] The poorer sections of the population were unable to purchase grain at the high rates that were prevailing in the scarcity-hit districts by the beginning of 1824 innumerable instances occurred of death from starvation and people were found to be selling their children into slavery.

The dislocation of agricultural operations affected the land revenue collections in the drought-hit districts. For instance, while Rs. 20,24,836 were collected during 1822-23 (Fasil 1232), only Rs. 11,29,324 could be collected in the next year in the North Arcot district.[27] Similarly the land revenue collected in Rayalaseema as a whole during 1822-23 (Fasil 1232) amounted to Rs. 44,63,926 while in the succeeding year it declined to Rs. 36,32,676.[28] During this year reduction in the assessment rates, as proposed by Munro was partially made in order to relieve the burden of taxation on the cultivators.

This was also the first occasion when a substantial amount of land revenue to the tune of Rs. 11,65,172 was remitted in the Rayalaseema districts. The Board of Directors of the Company approved the reduction of 25 percent on all dry and wetlands and 33 percent on lands depending solely on wells for the supply of water in the Bellary districts.[29] The collector of Cuddapah was reprimanded for burdening the cultivators with waste land and assessing them for land which they had not the means of tilling.[30] The Governor-in-Council ordered. The reduction of assessment should be general throughout the district and unfettered by conditions calculated to counteract its good effects. Even the Directors of the Company recognised the excessive and rigid character of the assessment and approved the proposal of the Madras Government to give relief to the cultivators in the ryotwari tracts in their present improvised situation.[31]

Great Storm of 1851

In May 1851 there occurred perhaps the most disastrous storm that the district had ever seen. On the afternoon of the 5th May heavy rain began to fall and it continued without intermission all that day and the next and the day after that as well. The storm swept from northwest to southeast across a tract of country about 50 miles in breadth. The collector wrote, "The damage done was excessive. Houses have been washed down everywhere. Cattle in hundreds and even thousands have perished in the rain." The roads, which lay in the track of the storm, were rendered impassable and 253 tanks within its influence in Bellary and Anantapur, having an ayacut assessed at 2.5 lakhs of rupees, were more or less seriously injured. Channels were swept away or obliged rated. much land was ruined by the sand which was deposited on it, the great tanks of Singanamala Anantapur and Dharmavaram were breached three-fourth of the town of Tadpatri were washed away and the great gopuram of temple on the river bank there was brought to the ground. The Singanamala tank was terribly damaged 840 yards of its embankment being swept away.

The estimates for the necessary repairs in Bellary and Anantapur amounted to three lakhs and as before it was ordered that only those works should be put in order, which would cost less than four years revenue to restore. An exception to this rule was however, made in the case of the Singanamala tank the cost of repairing which was over Rs. 60,000.

Bellary Famine of 1854

A severe famine, though limited in area, occurred in the districts of Bellary. Nellore and in the Western half of North Arcot in 1854. Other parts of Rayalaseema and the Circars escaped the rigours of famine in that year mainly because of the availability of employment to the vulnerable sections of IL, society as well as agricultural classes in the fields and the availability of food.[32]

The Causes

As on earlier occasions the famine of 1854 was preceded by a season of drought. The excessive rainfall of 1852 caused damage to standing crops in Bellary and elsewhere in the presidency. Around 35 inches of rainfall was recorded in Bellary, which was nearly double the usual average. Several tanks breached in consequence of the heavy rains and agricultural operation suffered. The following year turned out to be a disappointing one. The southwest monsoon was highly deficient in Bellary and in the taluks of Chittoor, Chandragiri, Palamaner, Venkatagiri Kota and Satyavedu of North Arcot district. Among the taluks in Bellary that suffered most mention may be made of Anantapur, Kalyandurg, Dharmavaram, Gooty, Adoni, Raidurg, Goolem and Penchapolium.[33] There was partial drought in Guntur, Krishna, Godavari and Visakhapatnam and famine was averted in these districts. There were abundant irrigation facilities to aid agriculture.

The poorer peasants and agricultural labourers in Bellary. North Arcot and Nellore due to the withering of the dry crops experienced acute distress. The scarcity of food contributed to the steep rise in prices.[34] Cholam, the staple food of the people raised from 58 seers a rupee to 27 seers in January 1854 in Bellary by June it raised to 21 seers a rupee. In the worst hit taluks of Kalyandurg, Dharmavaram, Gooty and Raidurg it was about 16 seers a rupee, conditions of scarcity deepened into severe famine on account of the capacity of the grain merchants. The collector of Bellary commenting on the grain dealers, wrote to the Board of revenue. Under these circumstances either the people must starve or by collecting in gangs for purposes of plunder under the pressure of hunger reduce the country to a state of anarchy which the local police cannot effectively prevent.

The small farmers were seen disposing of their livestock and immigrating to the neighbouring districts to eke out a living.[35] So the poverty-stricken peasants and labourers became the victims of hunger and disease. The weavers too felt the rigour of famine in Bellary district following the diminution in cotton production during 1853-54.[36] The difficulty of obtaining carriage facilities because of the scarcity of the beasts of burden enabled cotton merchants to enhance the price of raw cotton to nearly four times the usual one.

The basic causes of this famine were not far different from those of the famines discussed in the preceding paragraphs. As before, the extreme dependence on land due to the lack of other avenues of employment and the absence of dependable sources of irrigation (save the K.C. Canal). The continued burden of taxation compounding the general poverty of the agriculturists made them live on the verge of famine. The high prices, the agricultural produce fetched was not in any way helpful to the small ryots due to the reasons mentioned before. The condition of the landless agricultural labourers remained the same as in the sixties, their wages remained stationary while the prices rose.[37] The weavers and other village craftsmen, who eked out a meager living, also suffered from the effects of seasons of drought. The fixed income groups also were in hardship due to the soaring prices of food since 1875. The unscrupulous practices of the trading community such as hoarding, as also the continued adherence of the Government to Free Trade Policy, ably contributed to the distress of the people.

The seasons of normal rainfall between 1867 and 1873 indeed gave a spell of relief, but this came to an end in 1874 with the unusual floods of that year followed by years of adverse seasons, the long spell of dry weather since 1875 caused distress even to the larger landowners who gave up the old practice of storing grain as they were tempted by the lure of profits from the export trade thanks to the transport facilities.[38] Most of their profits, however, were spent on unproductive purposes.[39]

The Immediate Cause

The prelude to the famine of 1876 was the excessive rainfall of 1874 in Rayalaseema, Nellore and other parts of the Madras Presidency.[40] Nowhere in Rayalaseema the rainfall was below 35 inches, which 50 'id 100 percent above the normal. Bellary (including Anantapur), rated as the driest district in the Presidency, received a record rainfall of about 36 inches in 1874 prices of food began to increase rapidly since the end of 1875. Ragi, Cholam and Cumbu the staple grains of the mass of the population, registered a 200 percent increase between 1875 and 1876. The price of second sort rice also registered an unprecedented rise in 1876 as the following table of price show.

Famine of 1891-92

The southwest monsoon was deficient and cultivation was seriously retarded. Out of 12 lakh acres, only 8 lakh acres were sown. Even the yield of crops harvested was estimated to be less than half the average. Severe scarcity situation existed in Dharmavaram and Rayadurg taluks. Nearly one-third of the cattle in Anantapur district died. The review of the collector made soon after the famine showed its gravity the famine from which the district has just emerged was one of sorer trial to beasts than to man. It is impossible to say the mortality, but it must have been very great amongst the Jungle cattle.

Famine of 1896-97

The famine of 1896-97 happened to be one of the widespread famines of the century. It covered an area of 2,25,000 sq. miles covering a population of 62 millions in British India. As for Madras Presidency, it covered an area of 26,075 sq. miles and a population of 5.6 millions. In Andhra, besides Rayalaseema, the districts of Ganjam, Visakhapatnam, Godavari and Nellore were affected by the famine. It may be mentioned that the famine was not as severe in Madras Presidency as in other parts of India.

As usual, the famine had set in on account of bad seasons. In 1896 the southwest monsoon was erratic, which was followed by a near total failure of the northeast monsoon rains. The rains again failed in the months of June-July 1897. Consequently the price of food grains began to soar high. Between August and November 1896 there was a steep rise in the price of food grains. Cholam and ragi, the staple food of the people, registered a 100 percent increase, which made the condition of the average roots and the labourers miserable. The plight of the roots could have been mitigated partly atleast had they continued the old practice of storing grain for adverse seasons. But, thanks to the transport facilities available and the need for money, they sold away whatever surplus they were left with. They had not moreover expected the scourge to be at their doorsteps so soon after the last famine. So with practically no work in the fields and with no food grain stocks to tide over the adverse season, a large number of them and along with them the labourers, were driven to seek state relief even at the onset of the famine.[41]

It may be mentioned that, even though there was local scarcity, food was available in the market, in places where facilities for exploit of grain were not available. There were stocks of grain with the cultivators. For instance, at Dharmavaram in Anantapur district, which has a railway point, Ragi was sold at 19 seers a rupee in November 1896, whereas at Madakasira of the same district, which had no railway, it was sold at 30 seers a rupee. Partly on account of that there was enough food at the latter place to check the rise in price.[42] Further, on the average 30 wagons of grain was reported to have been passing through the Guntakal railway station daily for over three months to Anantapur and Bellary.[43] The private dealers in grain of Hindupur, Penukonda, Bellary, Cuddapah and other towns had sufficient stocks of grain. So grain can be had if only the need had the money to buy it. As usual the grain dealers and the well-to-do roots hoarded the grain in anticipation of high prices.[44] The Government too, for its part continued to maintain its hands off policy in food trade strengthened by the guidelines provided in the Famine Commission Report.[45]

Famine of 1896-97

There was water scarcity in Bukkapatnam, Dharmavaram, Anantapur, and Parigi tanks. Consequently, there was acute shortage of drinking water in the entire district. Almost one-fourth of the crop had failed. The prices of jowar and ragi shot up. The situation was not conducive for getting daily wage labour. Even the medium farmers were reduced to the position of wage labourers in their own farms and elsewhere. The areas worst affected were Tadipatri, Dharmavaram and the tract round Uravakonda.[46]

The famine condition was so critical that British government had arranged food grain from Mysore. There were two special trains from Hindupur to Guntakal for getting the food grains, which clearly indicate the gravity of the situation. Grain doles were also distributed in all the affected taluks. Scarcity of fodder need no special mention during this period. Forests also got dried up. The stalks of the faded crops and the roots called 'Sonti Verlu', scraped with much labour, supported the cattle for some time. Boda grass, normally used for thatching sheds, was in such great demand as fodder that even the hills were denuded of it. In Bordali firka of Hindupur taluk and in parts of Uravakonda and many other places, cattle were fed with prickly pear and with the leaves of neredu, yepi, palmyra and margosa. Fodder depots were established in Gooty, Guntakal, Anantapur, Dharmavaram and Chakralapalle with the help of the Famine Fund, Madras. According to one estimate, 9,200 cattle died. The then collector admitted that death of cattle could have been much more than what is revealed [47]

During this period of famine, private charity did its best to help the destitute population. The Indian Charitable Famine Relief Fund also spent in the district about a couple of lakhs of rupees towards relief. The intermittent showers that started towards the end of August slowly succeeded in wiping off the famine. The relief works were also closed by the end of October and "the last coolies to go, casting a longing lingering look behind, were the Yadiki and Uravakonda women and children". The more important famine works that executed were the Tadipatri-Yellanur road, Putlur-Tadipatri road, and repairs to Rayalacheruvu-Bhogasamudram road, Anantapur-Tadipatri road, Guntakal-Uravakonda near Uravakonda and the Demajipalle-Nayanipalle road.[48]

Anantapur, Kalyandurg, Gooty, Tadipatri taluks were worst affected due to this famine. Dry crops over large stretches either failed entirely or yielded far below the average. The prices of food grains had tremendously gone up during this year. Though 'tank repair works' were announced as part of relief measures by British Government, ultimately works did not get materialized. During 1901 famine period, Gooty and Tadipatri taluks were worst affected. Thousands of people were thrown out of employment in this particular year. Again in 1920-21 famine, due to failure of southwest monsoon, Gooty taluk became worst affected. Two test works and seven kitchens were opened in January 1921 in Gooty. The Indian People's Famine Trust made a grant from which clothes were supplied to destitute.[49]

Indian Irrigation Committee was formed in 1902 with Sir Colin C. Scott-Moncrieff as president, three British and one Indian as members of it. This committee report declared that "protecting Rayalaseema region from famines is the national duty".

Famine of 1924-25

Gooty, Madakasira, Penukonda, Dharmavarara, and Hindupur taluks were worst affected. Crops got dried up. There was no mark of seed germination in any of these regions. Madakasira taluk which was popular as 'Garden Area' had to face severe shortage of fodder. Consequently, all these taluks were declared by British government as famine-affected areas.[50]

Famine of 1934-35

Penukonda, Hindupur, and Tadipatri taluks were worst affected tank repairs, road works and other famine relief measures were taken up during this period. To relieve fodder scarcity, depots were opened in Penukonda, Hindupur, Kadiri, Dharmavaram, Anantapur, Tadipatri, Guntakal, Gooty, Kalyandurg and Madakasira. Among those organizing relief, the Rayalaseema Central Famine Relief Committee, the Indian Red Cross Society, the Indian People's Famine Trust, The Madras Corporation and the Zamindar of Ellamarri were the most prominent.

Famine of 1937-38

Gooty, Tadipatri, Hindupur, Rayadurg and part of Penukonda taluks were worst affected. The conditions in Gooty taluk and the Jutur area of Tadipatri taluk were so bad that even women belonging to some 'respectable' families were obliged to seek relief at the famine relief works. The weavers in Uravakonda were in straits and some of them were reported to have migrated to Bombay in search of livelihood. A relief centre had to be opened in the last week of June for providing work to them. Six relief works were also started by Government, two each in Gooty, Uravakonda and Tadipatri taluks and all these attracted a fairly large volume of labour from neighbouring districts. A depot was set up at Bantanahal in Alur taluk of Kurnool district to provide fodder for the Uravakonda area. Hill grass was sold at concessional rates and panchayat reserves were also thrown open for free grazing.[51]

Famine of 1941-42

The complete failure of the northeast monsoon of 1942 retarded agricultural operations and resulted in a failure of crops leading to conditions of scarcity. The entire district excepting the taluks of Kadiri and Hindupur was affected and relief measures covering nearly 28 percent of the district population had to be undertaken by government for almost a year. Clothes were donated from several other places to Anantapur district. Maharaja of Mysore and the Indian People's Famine Trust had sent donations. The Rayalaseema Famine Relief Committee also ran a few cheap grain depots.[52]

Famine of 1945-46

Bhogasamudram tank in Penukonda taluk, the main source of drinking water supply had completely dried up. People went with bowls to drink water at small scoopings made in its bed clearly indicates the water scarcity existing in at that period. 31,000 children were given free food by government. Over 6,300 lorries of hill-grass were imported to tide over the crisis which lasted till August, 1946, clothes including sarees were also distributed in the district.[53]

After Independence, the district as a whole was affected by droughts in 1951-52, 1952-53, 1957-58, 1960-61, 1962-63, 1965-68, 1968-69, 1970-71, 1975-76, 1976-77, 1977-78, 1978-79, 1979-80, 1980-81, 1982-83, 1983-84, 1984-85, 1986-87, 1990-91, and 1991-92 (details are presented in the table 2.3 given below).

Table 8: List of Famine, Drought and Scarcity Years and Cost of Relief Provided in Anantapur District (Rs. in Lakhs)

Year	*No. of villages affected*	*Cost of relief*	*Land Revenue remission granted*	*Taccavi loans advanced*	*Maximum % of population on relief*	*Remarks*
1390-91	—	—	—	—	—	Acute famine
1424	—	—	—	—	—	Acute famine
1792-93	—	—	—	—	—	Severe famine
1803	—	—	—	—	—	Near famine
1806-07	—	—	—	—	—	Scarcity
1823-24	—	—	—	—	—	Scarcity

contd...

Year	*No. of villages affected*	*Cost of relief*	*Land Revenue remission granted*	*Taccavi loans advanced*	*Maximum % of population on relief*	*Remarks*
1833	—	—			—	Severe famine
1838	—	—	—	—	—	Near famine
1853-54	—	—	—	—	—	Famine
1866	—	5.5	—	—	—	Famine
1876-78	—	20.2	5.38	5.23	7.98	Worst Famine
1891-92	—	0.16	2.64	2.34	0.38	Famine
1896-97	—	6.88	6.96	5.00	11.70	Severe famine
1900-01	—	0.19	1.35	0.09	0.30	Scarcity
1920-22		4.75	3.89	4.65	16.12	Famine
1924-25	—	1.70	1.59	5.00	2.70	Famine
1934-35	—	9.11	9.05	9.19	21.68	Severe famine
1937-38	—	5.04	5.34	0.94	18.10	Famine
1942-43	—	55.42	7.06	—	27.70	Famine
1945-46	—	14.25	8.79	1.67	1.60	Famine
1951-53	523	107.41	11.76	29.41	17.00	Severe drought
1957-58	388	92.30	4.65	—	—	Scarcity
1958-59	520	86.40	6.82	—	—	Drought
1959-60	684	38.50	7.72	—		Drought
1960-61	695	14.00	9.36	25.00	—	Drought

After Independence, there were two severe droughts, one in 1951-53 and other one in 1965-68.

This drought is popularly called as 'Ganji Karuvu'. During this period, agricultural labourers wandered in search of work. Purchasing power of farmers went down. The people of Kadiri, Anantapur and Penukonda taluks, started 'Karuvu yatralu' (drought rallies). In the process of subsistence, people of these taluks had to sell away cattle, ornaments and also kitchenware.[54]

According to the Commissioner of famines, in 1952 people of Kadiri, Penukonda, Kalyandurg, Dharmavaram, and Rayadurg taluks moved in atches in search of employment. He points out that, had there been no floods in 1952, drinking water problem in Anantapur district would have been far beyond imagination. In Kadiri taluk, people survived by consuming a leafy vegetable known as Ghatheraku. Realizing the gravity of the situation in the district, government involved military also in deepening the wells.

Moreover, in the region of Guntakal, Perur Upper Canal Project work was quickened. According to the records, there were 2.04 lakhs of people per day survived with 448 'Ganji Kendralu' (gruel centres). Four Ganji Kendralu were started exclusively for weavers in 1953. Due to heavy crowds at relief centres, cholera was intensely prevailing in the district.[55]

During this period, 650 tonnes of Rice, 1300 tonnes of wheat were imported from the then Soviet Union. Similarly, 2,436 bags of Rioe and 292 tonnes of milk powder were supplied by UNICEF. The Madras State Famine Relief Fund allotted 1.62 lakhs for the distribution of cloth, shark liver oil, medical aid and cash grants to pregnant women.

Much more severe drought conditions prevailed during 1965-68. This was a period of recurring droughts, affecting 156 districts in the country. In 1968, monsoon failed to provide sufficient protection to the Rayalaseema region of Andhra Pradesh. The year 1966-67 proved to be a turning point in the history of drought in India, affecting a large part of India. Two successive droughts, unprecedented in their spread and intensity, produced conditions of severe scarcity in 1966 and a famine like conditions in 1967 first since Independence. In Andhra Pradesh, drought affected about 130 lakh people spread over 17,340 villages in 130 taluks, among 65 taluks, drought continued for 3-4 years. Conditions were particularly dismal in Anantapur, Cuddapah and Kurnool in *Rayalaseema*, Nalgonda and Mahaboobnagar in Telangana and Srikakulam and the uplands of Guntur in the circar region (Singh 1978: 16-18).[56]

From the above accounts, it is clear that Anantapur is one of the chronically affected drought-prone region for ages. During the last 108 years, Anantapur district was affected by droughts as many as 52 times. That is, almost every alternate year had been a drought year affecting social, economic and other aspects of life. Moreover, it is not only a particular part of the district but entire district was under famine-drought conditions for a very long period of time.

Famine of 1951-53

Between 1948 and 1951 the district did not receive normal rainfall in either of the monsoons. With their total failure in 1951-52 it was once again the scene of a famine. The series of adverse seasons crippled the purchasing power of the ryots and agricultural labour started clamouring for employment. There were 'hunger marches' in Kadiri, the taluk worst affected and also in Penukonda and Anantapur. Agriculturists in these taluks were forced to sell their cattle, ornaments and utensils for their subsistence. In 1951, a slump in prices of some of the marketable products in Kalyandurg taluk posed a bigger problem than the one created by the adverse season. This resulted in a serious setback in the demand for labour and in the rates of wages. A famine Commissioner, appointed in 1952, found during his tours people wandering in search of employment particularly in Kadiri, Penukonda, Dharmavaram, Madakasira and parts of Tadpatri taluks. The only taluks comparatively free from the pangs of distress were Gooty and Uravakonda. Drinking water scarcity became acute in some of the worst affected tracts and but for the cyclonic storm in May 1952. The position would have been intolerable. These untimely rains, however, did not prove beneficial to agricultural operations with the result that the district lapsed again into a worse state of distress. In Kadiri taluk, in particular people were even driven to the necessity of consuming a leafy vegetable known as *ghatheraku*. Over 17 percent of the population was on relief. The rains in March and April 1953 were again unhelpful for agricultural operations. The distress was so intense that the sight of the milling crowds that gathered at the gruel centres visibly moved Pandit Nehru, who toured the district.

While the provisions of the Famine Code were not applied in carrying out relief measures, substantial relief was provided by means of works of lasting public utility. Low Scale Famine Relief Works, similar to the test works prescribed in the Code, were also started later with a view to ascertain with agricultural labour was willing to accept work at low wages. In addition to measures like liberal remission of land revenue, expansion of minor irrigation works. District board and departmental works, relief operations were undertaken on a very extensive scale.

To relieve scarcity of drinking water, deepening or wells was taken up under the Rural Water Supply Programme. A special subsidy scheme was also introduced temporarily in April 1952, for deepening both private and public wells, the 473 wells were deepened in the district, under this scheme up to the end of March 1953. Lorries of the Civil Supplies Department were diverted for transport of drinking water to the villages badly affected. The army also came to deepen the drinking water wells at Guntakal, a place proverbial for water scarcity, irrigation, rural water supply road, super-standard forest and rural welfare works were stepped up to provide labour to the unemployed. The Upper Pennar Project which was under execution and the canal excavation under the Perur Project, taken up as a relief work, also served to provide some employment.

Gruel centers were started initially from out of the amounts made available from the Prime Minister's National Relief Fund. As the distress mounted, many more were opened duly financed from State Funds and subsidized by the Union Government. Their number went up from 10 in February 1952 to 370 in August feeding over 1.15 lakhs of people everyday and rose to 448 in November with a daily attendance of little over 2.04 lakhs. As the relief extended to weavers through co-operatives was found inadequate, 4 gruel centers were opened separately for them in February 1953. A wide network of ration and fair price shops numbering 302 by the middle of the June also helped in meeting the situation.

Apart from the usual grazing concessions, paddy straw from the surplus Circar districts and hill-grass from the Nallamalas in Kurnool were imported and sold at subsidized rates so as to relieve scarcity of fodder.

A scheme of special remission more liberal than the role prescribed in the departmental manuals, was introduced in F. 1361 (1951-52). It was further revised in the following years so as to allow for more equitable treatment of tracts. According to the degree of distress suffered. Collection of land revenue and of cesses and loans was postponed. Subsistence loans to landholders, who had neither money nor grain and interest-free loans for the deepening of wells, were also liberal granted.

The public health measures undertaken included anticipatory anti-cholera inoculations to those attending the gruel centers and relief works, besides the free distribution of multi-vitamin tablets.

The Indian Red Cross Society and the Ramakrishna Mission rendered youngmen service. 650 tons of rice and 1.300 tons of wheat from Russia and 2.436 bags of rice and 292 barrels of milk powder from the UNICEF were among the more prominent of the gifts received from abroad. The Madras State Famine Relief Fund allotted Rs. 1.62 lakhs for the distribution of cloth, shark liver oil and medical aid and cash grants to pregnant women. Out of an allotment of Rs. 17,000 from the Madras Famine Relief Fund. the Guild of Service maintained 'work centres' at Gooty, Tadpatri, Anantapur and Hindupur to provide employment to poor and needy women.

Table 9 Indicated the expenditure incurred on relief during the various droughts and famines

Years of famine	*In lakhs of Rupees*			*Maximum percentage of population on relief*	*Highest monthly average number on relief*
	Cost of Relief	*Amount of Land Revenue remission granted*	*Amount of Takkavi loans granted*		
1876-78	20.17	5.38	5.23	7.98	1,68,384
1891-92	0.16	2.64	2.34	0.38	1,342
1896-97	6.88	6.96	5.00	11.70	84,893
1900-91	0.19	1.35	0.09	0.30	2,393
1920-22	4.75	3.89	4.65	16.12	21,636
19/4-25	1.70	1.59	5.00	2.70	9,424
1934-35	9.11	9.05	9.19	21.68	81,937
1937-38	5.04	5.30	0.94	18.10	32,009
1942-43	55.42	7.06	—	27.70	3,23,516
1945-46	14.25	8.79	1.67	1.60	19,056
1951-53	107.41	11.76	29.41	17.00	

The figures relate to the old Bellary district, which included Anantapur district.

Famine Relief Measures

The Government as before desired the district collectors to provide employment to the poor on public works more particularly on road works. They were given discretionary powers to offer this kind of relief without waiting for previous sanction in the event of the circumstances being emergent.[57] In the Bellary district over 97,000 people were employed on relief works during July 1854. Nearly 8 percent of the populations were on government relief at the height of the famine in this district. The Government spent Rs. 16 lakhs on famine relief in the district during 1854. Import duty of grain levied under section 22 Act VI of 1844 was suspended. The demand for grain in the famine-stricken districts was widely publicised in other parts of the country in order to attract imports. Messrs Remington and Ferguson and company, proprietors of Coffee Estates in Wayanad, proposed that they would employ 1,300 coolies from Bellary in their plantations provided that the expenses incurred for their travel was borne by the Collector of Bellary and that the Government would guarantee the Company from any loss in the venture. The Government declined the offer on the ground that there was no such precedent and it advised the company to operate through their own agents.

The Errors in the Relief Policy

The Government of Madras continued to adhere to the policy of non-interference in grain trade even under periods of extreme scarcity. For instance, the Collector of Nellore was not permitted to import grain on Government account. The Government also warned the Collector against interference in grain trade. He was ordered that on no account he should attempt to fix or regulate prices and that whether directly or indirectly by bringing Government into the market to compete with grain merchants. The hands off policy of the government helped in artificially raising the prices of grain beyond the reach of the poor in the scarcity-affected tracts.

Although definite statistics of mortality from this famine are not available, the population of Bellary declined in the quinquennium ending with 1856-57. The following table would give a comparative picture of the population (1851-52 to 1856-57) in the irrigated and non-irrigated districts to illustrate the effect of the famine of 1854 on them. It reveals that while the population had increased between 1851-52 and 1856-57 in the districts of Ganjam, Visakhapatnam, Godavari, Masulipatnam, Guntur, North Arcot, Nellore and Kurnool. It had declined in Bellary, the spurt in population in Nellore and the Circars may have been due partly to the migration of people from the affected districts of Rayalaseema.

Causes of Famine in Anantapur District

After making a brief survey of historical, socio-economical, agricultural, demographical, morphological and climatologically aspects, it is interesting to probe into the details of the causes for the recurrence of famine in Anantapur district. Occurrence of famine is a random event and it is difficult to establish a particular cause. Empirical evidence in the district of Anantapur shows that the forces behind famines of this area are more 'natural' than artificial or man-made ones.

Natural causes of famine generally include:

1. Droughts
2. Soil composition
3. Floods
4. Plant disease, destruction of crops by birds, pests, locusts, animals and
5. Irregular monsoons and hailstorms

Artificial causes are too many, in fact, artificial causes do not directly influence 'Famine' in Anantapur district. But this cannot be generalized, the role that artificial causes play is confined to the perpetuations of famine atmosphere. Therefore, in Anantapur district artificial causes are secondary in nature.

Also the artificial causes of famine involves three major sets of factors:

1. Background factors
2. Pathological factors (precursors)
3. Precipitating factors

Background factors	*Pathological factors (precursors)*	*Precipitating factors*
Basic biotic and ecological factors (eco systems)	Meteorological	Floods
	Biotic, Economic, Social (including political), etc.	Storms, Drought, War, Blight

In other way, the causes of famine can be stated as:

1. Food grain availability decline
2. Lack of purchasing power
3. Combination of a & b.

Various Indicators

The foregoing pages describe the chain of causation that lead to famine under different situation in the Anantapur district and indicate the possible scenario following famine which reflect the process of adjustment of the individual and the society to a post famine situation. In essence, one needs the combined study of the economy and the society to understand how these operate under normal circumstances:

(a) What kinds of shocks induce a chain of reactions leading to famine?

(b) What happens during a famine?

(c) Whether the economy and society adjust back to the original position or to a new position?

It is likely that under the prevailing circumstances quasi-famine situations (understood as widespread hunger and starvation) had affected an increasingly larger proportion of the populations of the Anantapur district. This is closely related to the phenomenon of "below poverty level equilibrium trap", which the district has experienced very often.

Prior to eighteenth century, in the district many famine death situations due to food grain intake shortage, i.e. due to starvation and hunger. This phenomenon in the district have given rise to number of symptoms involving individuals and the community as a whole which ultimately lead directly or indirectly, to excess deaths in the district. The symptoms are:

1. Increase in rural migration,
2. Increase in crime,
3. Increased incidence of fatal diseases,
4. Loss of body weight,
5. Eating of alternative "famine foods"
6. Changes in nutritional status,
7. Mental disorientation,
8. Uprooting of families,
9. Separation of families,
10. Transfer of assets,
11. "Wandering" and breakdown of traditional social bondage.

Among these, crime, disease, loss of both body weight, changes in nutritional status and eating of alternative "famine foods" combined together increased the death rate in the district in the medieval period. Even to this day the situation has not been changed very much.

Table 9 Land Use and Cropping Pattern

1	Total Geographical Area	19.13.400
2	Forest	1.91,677
3	Barren and uncultivable	1.75,676
4	Land under Non-Agri. Use	1,62,304
5	Permanent pastures and other grazing lands	23,342
6	Miscellaneous tree crops & groves	11.419
7	Culturable waste	68,692
8	Other fallow lands	1,09,645
9	Current fallow	2.00.203
10	Net area sown	9.67.535

Table 10 Important crops in the district

Sl.No.	Item	Kharif	Rabi
1	Groundnut	708283	22818
2	Jowar	11401	37711
3	Sunflower	7349	25164

Social and Political Factors

A detailed examination of the social and political aspects of Anantapur district substantiates the fact that in the permeation of the effects of drought, social and political factors do play an important role.

Social Factors

Among the social factors polarization of groups on caste basis is a predominate cause. Caste system has helped to divide the entire population of the district into a number of groups who control the entire means of production. The rationale behind their outlook is social prestige, caste system encourages a particular set of value system in which people waste lot of resources on the preservation of institution like marriage, family, rituals, etc. Caste system coupled kin and value system has contributed to the growth of social inequalities. Production takes place on the basis of individual needs and not on the basis of social needs.

Majority of the people (belonging to lower income group) have been addicted to alcohol and other intoxicated drinks. Ignorance is very much prevalent partly due to illiteracy and partly due to the non-exposure to the urban centers. Laziness and feudalistic attitude are widely observed in Anantapur district which breeds envy, jealousy, and intolerance on the one-side and lack of means of livelihood on the other side. Comparatively in Anantapur district the incidence of thefts, robberies and insecurity of property, etc. have been high.

Given the system of production based on individual needs, the quantum of marketable surpluses in food grains is not appreciably high. In the event of an inappropriate or shortage of rainfall, the surplus food grains are not sufficient to meet the minimum needs of this region. Purchasing power of the masses is inadequate due to low levels of income. Hence large-scale starvation is seen as richest manifestation of drought.

Rural people in Anantapur district, who constitute the bulk, do not possess social awareness. Ignorance, superstition, illiteracy and the consequent take it for granted attitude come in the way of increasing the severity of famine. When people are educated and enlightened, they take initiative in chalking out and implementing plans for integrated rural development that relieve them from food scarcity, hunger, unemployment and poverty. But there appears a tremendous change in the social environment in the past decade in the district. The following table shows the improvement in literacy rate.

Socio-economic Factors

The detailed picture of socio-economic effects of famine including starvation, mortality, slavery and cannibalism during the early centuries has been left by contemporary chronicles. European factory records and traveler's accounts. These may be analysed into the following categories:

(a) Food shortage and rise in prices

(b) Break-up of family life, sale of women and children and voluntary enslavement

(c) Cannibalism

(d) Migration

(e) Pestilence

(f) Death & depopulation

(g) Ruining of agriculture

(h) Paralysis of industries

(i) Effects on marketing conditions

(j) Disruption of inland transport and trade

(k) Dislocation of foreign trade

(l) Lasting effects, if any

It is possible to work out these effects with reference to individual famines. But perhaps most of these were illustrated during the "dun kuruvu" of 1876-78, the most devastating and the most representative of all the known calamities of the period in the district. While the loss of life and intensity of human misery, leading to deaths from starvation, suicide, the break-up of family life aimless wandering, voluntary enslavement and sale of children will primarily interest the social historian and these also had a significant bearing on economic life.

Political Factors

People in the district are politically less matured than the people of other region of Andhra Pradesh. But the working of the system of Panchayat is such that the politicians in power are dominating the scene. At the village level the intrusion of politics in day-to-day life has influenced disintegrating forces into the set up. It is generally difficult to find villages united to fight famine condition. At the state level, politicians and people's representatives develop interest in particular places only. The tendency to dominate and block administrative machinery on the part of the politicians and the all pervading selfishness of the people in power generally ignore the principle area and regional planning and conceive of unscientific plans, resulting in lop-sided development, political willingness to dump more finances and political leadership to identify and promote growth enters in backward and frequently drought prone and famine affected areas must come forward to reduce the undesirable effects of famine.

Economic Factors

Economic conditions of Anantapur district have a close bearing on famines (History of famine in India reveals the fact that during the colonial rule in India there has been an increase in the frequency of famines due to the deteriorated economic condition of the masses on account of making India too much dependent on agriculture and neglecting diversification of economic activities). The frequent recurrence and severity of famines in the district during the tenth century is due to economic conditions.

The following are some of the causes of famine in Anantapur district:

1. Lack of irrigational facilities
2. Under developed transport, communication and power
3. Cumulative nature of famines
4. Excessive dependency on Agriculture
5. Small scale farming and defective land utilization
6. Worklessness in rural areas (due to population explosion)

Lack of Irrigational Facilities

Although Andhra Pradesh is a 'River State' famous for its water resources. Anantapur district a component of Andhra Pradesh is not endowed with similar water potential because river waters are not brought to the district for irrigational purposes. There are only two rivers, Pennar and Chitravathi in the entire district. The river themselves do not have water round the year. Only in rainy season the rivers flow with water and sometimes excessively to inundate the borders of the regions in which the river flow.[58]

The other areas of the district to a great extent depend upon canals, tanks and wells. This network depends purely on rainfall, which is not regular and sufficient in the district, hence the nearness of the district to drought calamities leading to famine. The nature of the major fraction of land is such that it cannot preserve moisture. This adds fuel to the fire, in the event of a shortage of rainfall, the poor network of irrigational facilities, fail to store water to contain the disastrous effects of famine.

The water resources of this district are less as compared to the other districts of coastal Andhra and Telangana, due to the lack of perennial rivers and also due to the erratic precipitation of the district.

The ground water is the only source of assured irrigation in drought prone backward areas with affordable investments and quick turnaround time without longer gestation periods unlike major irrigation projects. The major source of recharge to ground water is rainfall which constitutes about 8 to 15 percent. The rapid increase in ground water exploitation has resulted in fall in water table level and drying up of shallow dug wells/bore wells rendering themselves useless and leading to untold miseries to the farmers. Hence, there is an urgent need for the augmented rain water harvesting to increase the recharge from 10 to 15 percent to at least to 30 percent of the water resource availability in Anantapur has been calculated basing on the rainfall and run-off characteristics. The model is as follows:

Table 10 Water Harvesting in Anantapur District

Geographical	1913492 Ha
Average rainfall	520mm
Total precipitation over geographical area	9950 MCM
Ground Water recharge (village-wise estimate 11% of rainfall)	1101 MCM
Evapotranspiration	5970MCM
0 Run-off	2985MCM
Surface Water Storage	839.57MCM
Balance Run-off (f-g)	2145MCM
Potential rainwater harvesting (50% of h)	1072 MCM
g) Proposed harvesting through	
(1) Percolation tank(0.28MCM (x) 4 fillings (x) 0.6 (x) 2640)	177.41 MCM
Farm Ponds (0.003 (x) 2 (x) 289000)	(1) 173.40MCM
Check dams (0.014 (x) 4 (x) 0.6 (x) 1692)	56.85MCM
(1) Total(37 percent over present ground water recharge)	407.66 MCM

Now we understand as to how the district of Anantapur is more backward in respect of communication facilities, transport, power and the lack of these required facilities, constitute the causes of famine in the district.

Cumulative Nature of Famines

With frequent occurrence of famines the farmer's wealth is damaged by short rations and made them incompetent physically to cultivate the usual acreage for the next crop. Next crop becomes again smaller than the usual and things go from bad to worse. Although modern developments have made the incidence and the severity of the crop season less serious, the problem is still an alarming one in the district. Other ways in which famine tends to breed famine are seen when seed stocks or draught animals are consumed for food, thus making the securing of the next seasons harvest a still more difficult task. For examples in Anantapur district in the present century there were famines in 1921-23, 1924, 1926, 1931, 1934, 1937, 1942, 1945, 1950, 1951, 1952, giving only a time interval of two-three years in most cases between the visitation of one catastrophe and the other. The incidence of famine helps the creation of near-famine conditions. This is brought about by means of rural indebtedness. Indebtedness impoverished the agriculturists in two ways: Annually it deprived him of a considerable part of his income and very often even obliged him to sell the whole produce to the creditor at below the market rate: in the long run, it forced him to part with his land and reduce himself to the position of a tenant.

Excessive Dependence on Agriculture

Agriculture is the only activity, which directly depends on rainfall and irrigation facilities. It has been shown that the percentage and level of rainfall in Anantapur district is below the state average and is insufficient to meet the requirements in agriculture. Rainfall is a random phenomenon, which may also be untimely. This erratic nature of rain hampers food production and agricultural growth.

The below table 9 shows clearly that the dependents on agriculture constitute 75 percent of the population. The workers depending on non-agricultural sector are very low in number. This proves that the excessive dependency on agriculture is resulting in famine.

Table 11 Occupational Distribution of Workers, Anantapur

Sl.No.	*Classification of workers*	*1971*	*1991*
1	Cultivators	3,26,992	4,62,992
2	Agricultural Labour	3,46,090	4,89,286
3	Dependents on agriculture	6,73,082	8,76,089
4	Mining. Quarrying, etc.	27,575	42,095
5	Household industry	45,197	86,179
6	Trade & Commerce	41,827	2,38,127
7	Transport. Storage and Communications	16,237	96,932
8	Other Services	52,348	1,59,348

Source: *Statistical Abstract of Andhra Pradesh, 1976, pp. 26-27. Data collected from C.P. Office, Anantapur.*

Small Scale Farming and Defective Land Utilization

Small scale farming reduces the production and productivity of agriculture. Sufficient mechanization is not introduced in agricultural operations of this district. Defective land utilization in the district is due to factors like lack of water, lack of drainage, deep-rooted grasses and weeds, low fertility of the soil and lack of capital. Irrigational canals involve heavy capital outlay and cannot be built in a day. Land observed by jungle cannot be brought under cultivations without mechanization and capital. In dry cultivations, no green manure or chemical fertilizers are applied. The farm yard manure available with the cultivators is utilized mostly for wet crops and only once in a while for dry crops. Chemical fertilizers have not yet become popular due to the limitation set soil erosion in the district. All these factors are contributing to the low production and production of agriculture. The defective land utilization can also be seen in the district.

The above-referred figure reveals that the area of forestland is more in this district, comparatively to the other districts of Andhra Pradesh. The net area sown is very low. The pressure of population on the land is very great and the area of cultivable but uncultivated land, which can he brought under cultivation without the application of special effort is small.

Worklessness in Rural Areas

With increasing numbers added to the work force on account of the rise in population, the problem of unemployment, underemployment and disguised unemployment in the region are also causing to increase the intensity of famines. Anantapur district, with practically no industrial base in its economy and the agriculture, subject to chronic drought is a region where the problem of unemployment gets accentuated. While this is so the existing data are inadequate for building up a sufficiently detailed picture of the state of unemployment for the region as a whole. No doubt, it is not easy to measure the extent of unemployment in a backward district like Anantapur, for there is bound to be tendency, especially among the self-employed, to share work with the available work force.

As more than 75 percent of the population in this district is depending on agriculture the extent of under employment, which is chronic in the agricultural sector can very well be understood. A more significant feature of the agricultural economy in the Anantapur district is the predominance of dry cultivation where employment opportunities are much less than in wet cultivation. Agricultural labourers constitute the major section of the rural work force. Most of them being landless are depending upon the wage paid employment in agriculture. Their wages are low which do not permit them even to meet the bare needs of the subsistence. In the periods of unemployment they have to suffer from hunger. So they have to depend on the prosperous fanners and are becoming indebted which leads to the exploitation of these unfortunate sections. Thus, the rural unemployment has become one of the causes of famine in the dry land cultivated area of Anantapur district. B.M. Bhatia rightly pointed out that it is in this background of poverty of the agricultural masses, growing distributions among them, rising price of food grains, stationary wages and the increasing population obtaining little scope for the employment outside agriculture, that a series of famine occurred in the country. This situation is more and also applicable to Anantapur district.

Conclusions

Natural and artificial causes do play their part, through not equally, in the recurrence and perpetuation of famines in Anantapur district. Among the natural causes:

(1) Drought is the predominant one.

(2) Due to peculiar geographical position of Anantapur district, the average annual rainfall is below the state average and is within the range of drought susceptibility.

(3) The next important natural cause that leads to famine conditions is the dominance of red soil in the district. This soil by nature is incapable of preserving moisture.

Social political atmosphere in Anantapur district is such that famine is not understood in true perspective and seriously enough to the desired extent. Political leadership has not been bestowing sufficient willingness to take measures to alleviate the severity of famine to benefit the general masses in the district. Anantapur district has an underdeveloped network of canal system. Since canals are not linked to rivers and the water source for tanks is ultimately the rainfall. This region lacks irrigation facilities. This is one of the chief causes for low productivity and production of agriculture. The other factors are the general poverty of the masses. Lack of diversification in economic activities and lack of alternative occupation. Another important reason is lack of well-developed transport, communication and power-generation system. Social, political and economic causes do not directly influence famines. When once the famine comes, economic forces play a role to aggravate the famine situation and to prolong its continuation.[59]

Notes and References

1. Pasala Sudhakar, 'Drought Distress Suicide of Farmers in Anantapur district', In P. Prasangi (ed.), *Osmania Journal of Historical Research Bi-Annual Journal*, Vol. VII, 2006, No. 6, pp. 20-37.

2. The basic information on the district is drawn largely from the district's official website, Ministry of Irrigation and Power, 1972, p. 157.

3. M.V. Nadikarni, *Socio-economic Conditions in Drought-Prone Areas*, Concept Publishing Co., New Delhi, 1985, p. 24.

4. K.W. Olsen, 'Man-made Drought in Rayalaseema', *Economic and Political Weekly*, 1987, Vol. XXII, No. 11, pp. 441-43.

5. T. Scarlet Epstein, *Economic Development and Social Change in South India*, Oxford University Press, Bombay, 1962.
6. Paragraph 323 of the letter, dated 3.1.1825 of the Board of Revenue.
7. Quoted in W. Francis, *Anantapur District Gazetteer*, pp. 77-78.
8. Munro to board of revenue, 7th August 1804, Sir Thomas Munro's minutes, letters and dispatches on food scarcity in Arbuthnot, major-General Sir Thomas Munro, Vali, p. 214.
9. Revenue dispatches to England, 1821-1826, Vol. 60088 (T.N.A.), p. 163.
10. Near the eastern frontier of Cuddapah district.
11. Munro to Board of Revenue. 11th January 1805, Sir Thomas Munro 's Minutes, Letters and Dispatches on Food scarcity in Arbuthnot, Major-General Sir Thomas Munro, Vol. 11, p. 215.
12. Munro to Board of Revenue, 24th May 1807, Proceedings of the Board of Revenue, Miscellaneous Records, Col. Munro's proceedings March-October 1807, Vo1, No. 1805.
13. Munro to Board of Revenue, 20th August 1807, D. Maclean, *Manual of the Administration of the Madras Presidency*, Vol. I, pp. 299-300.
14. Munro to Board of Revenue, 9th February 1807, Arbuthnot, Major-General, Sir Thomas Munro, Vol. 1, p. 222.
15. *Ibid.*
16. *Ibid.*
17. *Ibid.*
18. General Reports, Vol. l, p. 195 (M.S.S. T.N.A.).
19. Proceedings of the Board of Revenue Miscellaneous, Vol. Bundle No. 17(7), pp. 13-15, 26th January 1807.
20. C.D. Macleane, *op. cit.*, p. 300.
21. *Ibid.*
22. A charitable Institution established by private subscription in 1782 to afford relief the famine-stricken people.

23. While the number of deaths recorded in Madras was 3,225 in 1805, it rose to 4,902 in 1806 and to 17,207 in 1807, R.A. Dalyell, *Memorandum on the Madras Famine of 1866.*

24. Secretary Board of Revenue to all the Collectors, 16th November 1812. Madras famine of 1866, Madras, 1867, pp. 24-25. quoted in A. Sarada Raju, *Economic conditions in the Madras Presidency 1800.*

25. Minute of Sir Thomas Munro, 12 December 1823 and Dispatches of Sir Thomas Munro, Arbuthnot, Major-General Sir Thomas Munro, Vol. 11, p. 228.

26. Secretary, Board of Revenue to the Government of Madras, 22nd December 1823, No.513, Revenue Consultations, 6th January 1824 (M.S.S. T.N.A.).

27. R.A. Dalyell, *op. cit.*, p. 31.

28. Collector of North Arcot to the Board of Revenue, 18th January 1833. General Record, Proceedings of the Board of Revenue, Back No. 40, p. 932.

29. Secretary, Board of Revenue to Government of Madras 22" September 1831, no. 416 Revenue Consultations, 11th October 1831, p.3412.

30. Letter from the Court of Directors to the Government of Madras, 18 August 1824, Dispatches from England (Revenue), 1823-1827, Gen. Nos. 6144-6145.

31. Proceedings of the Board of Revenue, 7th May 1824, No. 311, p. 1218.

32. C.D. Macleane, *op. cit.*, Vol. 1, p. 299.

33. W. Francis, *op. cit.*, p. 80: W.W. Hunter, *The Imperial Gazetteer of India*, Vol. 1, London, 1885, p. 246.

34. Proceedings of the Board of Revenue, 5 January 1854, back nos. 51-68.

35. Minutes of Revenue Consultation, 3rd January 1854, No. 3 in Proceedings of the Board of Revenue, 9th January 1854.

36. C. Melly to the Board of Revenue, 29th December 1853.

37. C.B. Macleane, *Standing Information Regarding the Administration of the Madras Presidency,* Madras, 1979, pp. 4-5.
38. *Proceedings of the Board of Revenue,* 7th September and 24th October 1876.
39. W. Francis, *op. cit.,* p. 54.
40. *Indian Famine Commission of 1898,* Appendix, Vol. II.
41. Besides the Madras Presidency, parts of united provinces, central provinces, Gujarat, Bombay, Punjab and Burma were affected by this famine. See *Report of the Indian Famine Commission 1898,* p. 176.
42. *Ibid.*
43. *Ibid.,* also see H.S. Srivastava, *History of Indian Famine,* pp. 199-200.
44. *Ibid.,* Vol. II, p. 3.
45. *Ibid..* see also H.S. Srivastava, *op. cit.,* pp. 201-202.
46. *Ibid.,* p. 286.
47. *Ibid.,* p. 301.
48. *Ibid.,* p. 287.
49. *Andhra Pradesh District Gazetteers,* Anantapur, 1970, p. 259.
50. *Ibid.,* p. 259.
51. *Ibid.,* p. 260.
52. *Ibid.,* p. 261.
53. *Ibid.,* p. 262.
54. *Ibid.,* pp. 262-263.
55. *Ibid.,* p. 264.
56. Tapeshwar Singh, *Drought-Prone Areas in India,* People's Publishing House, Delhi, 1978.
57. R.A. Dalyell, *op. cit.,* p. 45.
58. B.M. Bhatia, *Famines in India, 1860-1945,* p. 155.
59. *Ibid.,* p. 26.

Chapter VI

CONCLUSION

India's success in preventing natural disasters from developing into large-scale famines since Independence is not a spurious one. The entitlement system defined by the operation of the economy and the ordinary level of State provisioning leaves a large part of the population highly vulnerable to entitlement failures in times of crisis. On several occasions, famine would undoubtedly have occurred in the absence of early and effective intervention to protect the entitlements of vulnerable groups. If the Government of India can and should be criticised for having gone little further than espousing the earlier colonial view.

Two conspicuous milestones characterize the historical development of India's relief system in its modern form. The first one corresponds to the submission of the Famine Commission Report of 1880, and the simultaneous advent of the Famine Codes. The Famine Commission of 1880 saw the loss of employment and wages for agricultural labourers and artisans during droughts as the primary cause of famines, while it pronounced that food was rarely in short supply for the country as a whole. Accordingly, the famine relief strategy embodied in the Famine Codes consisted of generating purchasing power in affected areas, and letting private trade ensure the physical supply of food. Moreover, the preferred income transfer mechanism consisted of open-ended public works supplemented by gratuitous relief for the weak. The self-selection future of relief works was relied upon to ensure financial economy while providing a universal guarantee against starvation.

It would be simplistic, however, to regard today's relief system as a mere legacy of the British Administration. In fact, important changes have occurred since Independence, which marks the second milestone of the recent history of famine relief in India. In particular, the real resources allocated to relief have increased, and the range of measures they support has broadened; the Government has resolutely entered the previously sacrosanct domain of food supply management, and ensured a large measure of price stability and, last but not least, the real but weak commitment of the British Administration to the prevention of famines has evolved into a political compulsion to respond as a democratic institutions buttressed public pressure.

Against the background of a dramatic and prolonged collapse of agricultural production and food availability, the massive provision of employment succeeded not only in stimulating private food trade in a situation where the Public Distribution System had proved inadequate to the task of filling the initial gap between availability and requirements, but also in ensuring that the remaining deficit was distributed with astonishing evenness across different socio-economic groups. Accordingly, this devastating drought only had moderate consequences in terms of asset depletion, nutritional damage and excess mortality. The case study also highlighted the crucial role played (inter alia) by public pressure, cash relief and public works in averting a dramatic crisis. Public pressure galvanized the Government into prompt and decided action. Cash relief enabled the logistical resources of the public sector to be supplemented with those of the private sector, and insulated income support strategies from food delivery failures. Last but not least, the reliance on public works as the main income transfer mechanism ensured both a sharp concentration of resources on the needy and perhaps even more importantly, the provision of a nearly universal protection against starvation.

Only preliminary reflections have been offered here on the question of whether the Indian experience conveys important lessons for other countries, especially those of Africa. Many complex issues are involved, some of which can only be tackled on the basis of sustained empirical investigation. Meanwhile it remains pertinent to ask whether, in spite of the political obstacles one might expect, significant improvements could not be achieved in many African countries by (i) paying greater attention to the persistent need for a relief system in countries where a large part of the population.

Derives its subsistence from relatively unrewarding and insecure activities; (ii) acknowledging the vital importance of contingency planning for effective relief; (iii) concentrating on the anticipation, recognition and cure of sharp declines in living standards, rather than on the mere prevention of starvation deaths; (iv) giving a greater role to the local administration, and studying the incentives which can make it work; (v) working separately on the; support of incomes and the delivery of food, rather than making the former contingent upon, and commensurate to, the latter; (vi) considering more sympathetically the potential of cash relief; and (vii) taking a more flexible view of the possible role of private trade.

Drought is beyond question, among the more serious problems this country faces. Drought relief, almost equally beyond question, is rural India's biggest growth industry. Often, there is little relation between the two relief can go to regions that get lots of rainfall. Even where it goes to scarcity areas, those most in need seldom benefit from it. The poor in such regions understand this. That's why some of them call sought relief measures (the third crop) only, they are not the arises who harvest it.

A great deal of drought 'relief goes into contracts handed over to private parties. These are to lay roads, dig wells, send out water tankers, build bridges, repair tanks the works. Think rat can't total up to much? Think again. The money that goes into this industry in a single year can make the withdrawals from man's animal husbandry department look like so many minor fiddles. And the Bihar scam lasted a decade and a half. The tragedy, of course, is that it rarely addresses the real problems of drought and water scarcity.

In 1994-95 alone, the rich state of Maharashtra spent over Rs. 1,170 crores on emergency measures in combating drought and on other water- related problems. This was more than the combined profits the previous year of leading companies all across the country in the organised sector of the tea and coffee, cement and automobile industries. Their profits after tax came Rs.1,149 crores, according to a report of the Centre for Monitoring the Indian Economy.

In August 1995, Prime Minister Narasimha Rao inaugurated. anti-drought project in Orissa. This one will involve spending Rs 4,557 crores in six years (over Rs.750 crores a year) on just few districts including Kalahandi, Bolangir and Koraput.

In theory, drought-prone blocks come under a central scheme known as the Drought-Prone Areas Programme (DPAP). But bringing blocks into the DPAP is now a pure political decision. The central allocation for DPAP may be nominal. But once a block is under DPAP, a phalanx of other schemes follows bringing in huge sums of money. The same blocks then get money coming in under the employment assurance scheme (EAS), anti-desertification projects, drinking, water missions and a host of other schemes. Well, some people do benefit.

All this has happened during a period where there have been - several successive good monsoons. There has been scarcity too for some people. But that's a different story.

Kalahandi's major problem, as the reports in this section show, does not arise from poor rainfall. Water resources experts and administrators would largely agree that, barring problems of erratic timing and spread, most Indian districts could get by on around 800 mm of rainfall annually. The lowest rainfall Kalahandi has had in the past twenty years was 978 mm. That is way what some districts get in 'normal' years. Otherwise, Kalahandi's annual rainfall has been, on an average, 1,250 mm. That is pretty decent in 1990-91, the district had 2,247 mm of rainfall. Besides, Kalahandi produces more food per person than both Orissa and India as a whole do. Nuapada, the worst part of old Kalahandi and now a separate district, got 2,366 mm of rainfall in 1994.

In Palamuru, too, average rainfall is not bad. The district gets 200-1,230 mm of rain in a normal year. In its worst year in recent history, it received 630 mm. Some districts in India get less without experiencing the same damage.

Once it was clear that drought and DPAP were linked to fund flows in a big way, it followed that everyone wanted their block under the scheme. In many cases, the powerful are not only able to bring their blocks under it, but appropriate any benefits that follow.

Famines have occurred throughout history and routinely result in substantial death and destruction. However, a famine is not just a single event but rather results from a sequence of events and processes that reduces the availability of food generally or the access to food of particular vulnerable groups. This reduction in food may be either chronic or temporary and may be caused by both natural and human made factors. Among the human made factors, conflict is often a significant cause of famines and the presence of conflict frequently renders food assistance difficult or, even, impossible.

In theory, famines are preventable in reality, they occur all too often. When they do occur they represent a breakdown in the capacity of local, national and international assistance systems. "Victims" of famine must often rely first on their own resources to alleviate famine impacts. The design of relief programs must take account of these coping patterns and build on community and individual strengths. Sometimes these coping patterns are systematic enough to provide forecasts of developing famines. In addition to analysis of coping patterns, many countries now have "early warning systems" that measure drought conditions, food production and food marketing to alert authorities to an impending crisis.

Many different programs which help to maintain food security have proven successful. These include the distribution of food to general or targeted populations, income producing programs and price control programs to help households purchase food, complementary water and health programs to minimize morbidity and mortality, and special programs to maintain livestock and other household assets.

Effective government response is critical to successful relief programs in response to famine. Unfortunately, in many countries that experience famine governments are unstable thus inhibiting effective responses. Several United Nations agencies have important roles to play in responding to famine. Because many emergencies include, in addition to famine and displaced populations, civil conflict and intense political considerations, the UN is developing new roles to support effective responses to provide humanitarian aid. Increasingly, donors are choosing to deliver famine assistance through

NGOs. This trend has some significant implications for relief and rehabilitation programs.

Famines and droughts, starvation and migration have become part of the lives of people over the years in many parts of the world, particularly in the third world countries. Hence, there is a growing need for research in these regions for understanding social processes at large, and adaptive mechanisms that people have evolved in tiding over crisis situations.

Drought had induced farmers to think of innovations in the cropping pattern and accordingly farming practices, are well adapted to suit the harsh environment. For instance, crops with different maturation periods are cultivated in rotation to cope with erratic rainfall. Similarly, mixed cropping and inter-cropping are also adaptations to the prevailing conditions. The knowledge of seasonal migrants also helped in shifting from one variety to another variety of grain, which gives a higher yield as an adaptation the local environmental conditions. This was evident with the shift from local variety of jowar to the Raichur jowar, which is a high yielding variety, that were brought by seasonal migrants as part of their wage labour.

Patron-client relationship acts as buffer, to a certain extent, during drought periods. Though land lease, credit giving etc, between patron and client, apparently seem to be economic transactions, have a lot of social significance underlying it. Adjusting consumption of food is an important adaptation during drought years. But, the adaptations among big and medium farmers are quite different from that of small, marginal and landless categories. In case of later, adjustment involves reduction in number of meals taken in a day (from three meals to two meals a day), solid food to liquid food, shift to coarse grains, and eating various kinds of forest leaves. This being so, medium and big farmers resort to reduced consumption of milk, pulses, sugar, vegetables and food grains. In addition, liquidating assets, resorting to share cropping, borrowing for consumption as well as production are the other commonly observed adaptations. For additional support, households turn to jati and kin members, neighbours and hereditary patrons.

Not meeting certain obligations like gift giving to sons-in-law are considered as socially legitimate during drought period. Several socially and culturally important occasions also are deferred or postponed. Social visits and expenditures are curtailed to a large extent Some times, people are forced to violate the existing customs while tiding over the crisis situations, which ultimately have lasting impact in their living pattern. For instance, among the Madiga jati, the obligation to make available abundant liquor to all people present on the occasion of marriage became so much a liability on part of the bride groom party during droughts that the jati council ultimately decided to abandon liquor giving.

Several adaptations that have taken place in the study village over a period of time have both short-term and long-term implications for the people. The implications include permanent migration, increase in the area of land under cultivation, increased leasing activity among seasonal migrants, significant change in land transactions, lower strata of people organizing themselves into pressure groups like Rytu coolie sangam, jati groups like valmiki sangam, etc. effective utilization of CPR's and increase in non-farm activities. The other implications of seasonal migration include that of children abandoning schools and consequently discontinuing education, vulnerability to ill-health and accidents at the place of migration, leading sometimes even to death and suffering for the dependents.

A clear distinction could be observed between the relatively rich and poor in resorting to sale of assets. Big and medium farmers try to sell the livestock during droughts because of non-availability of fodder, while small and marginal farmers are forced to resort to the means or selling draught animals to meet their consumption needs as well as to get little money that could be utilized for seasonal migration. Though cultivator jati households have purchased more amount of land per capita than any other jati households in the village, it is also true that the purchasing power of the agricultural labour jati and Scheduled jati households has increased. One of the reasons being that the seasonal migration has given enough exposure to use land more productively.

Change in the hereditary relationships and diversification of traditional occupations is an important consequence in the process of coping up with drought. For instance, in case of the carpenter household, a hereditary relation was replaced by a contractual relation. This is because farmers have adopted agricultural implements made of iron, instead of traditional wood-made ones. This shift in implements made the carpenter household to diversify the occupation into making windows, doors, tables, etc. Similarly, Kummari jati household diversified its occupation into making earthenware materials such as toys which are commercial value.

Seasonal migration has its impact on the social fabric of the village. At the place of migration, a few married women and unmarried girls eloped, unable to cope with the situation and subsequently, families got divided down giving severe jolt to the existing value system in the study village. A few households laid to resort to the worst means of living by begging. Though Madiga jati people resorted to this initially, later on even other jati members, whose social status was high also took recourse to it.

Women play a crucial role in dealing with droughts. In fact, they adapt well according to the household needs and also suggest adjustment mechanisms that are quite suited to the households in the periods of drought. This is evident in terms of negotiating small amounts of credit, maintaining social bonds and kin relations, generating additional source of income, offering themselves in agricultural and non-agricultural activities, along with responsibility of domestic work and child care. In terms of stress and strain caused due to drought, women are the most affected. They undergo sexual harassment at the places of migration, receive insufficient care during pregnancy and are at a disadvantage while adjusting consumption patterns.

The beliefs on droughts are important in the sense that they reflect deep conventions which are being followed from age old times to this day. For example, the uncertain and erratic rainfall has created a kind of fear that exists always in the minds of people. Since droughts are seen as resulting from intervention of a divine force, it is obvious that people have evolved methods of appeasing these forces in their own ways. The rituals that are practiced in the study village to appease the rain god include exhuming the buried people, hundred and one pot ceremony, Banalu velladam along with a few other rituals seeking rains. Moreover, these practices indicate the anxiety of farmers regarding prospects of rains and good crop years.

Many government programmes are reaching the village with varying coverage and impact. These include subsidized credit, Public Distribution System (PDS), Land distribution and few other schemes like sheep, bullock and bullock carts distribution. The government programmes that take drought relief measures into consideration are IRDP, JRY, DPAP and EWACRA schemes at district, mandal and village levels. Postponement of land revenue collection and, opening of free fodder centres are other governmental measures to mitigate the drought conditions. NGO's play a critical role in making people realize these schemes better by way of increasing their awareness. In the study area RDT (a voluntary organization) has taken up various developmental schemes keeping the thrust on education, health, women and ecology Rayalaseema in general Anantapur district in particular.

Both governmental and NGO's have been contributing to the development of drought-affected people. However, one does not notice any significant impact on the living patterns in the study village. Compared to governmental programmes, the impact of ROT programmes on the people is more visible. This is due to the fact that, unlike the government agencies, the RDT personnel are in constant touch with the people. Though government policies intend to help the poorest among the rural poor, the actual process of implementation does not match accordingly. Needs and priorities of local people should be taken into consideration in implementing both short-term as well as long-terra measures, in backward areas like Anantapur.

The study illustrates the overriding importance of seasonal migration and the role of external influences in adjusting to sharp drought-induced shortfalls in income. The role of seasonal migratory process in conditioning human welfare in times of crisis has been clearly established Breman. However, seasonal migration is not merely confined to the landless, marginal and small farmers but extends even to the medium and partially Dig farmers during severe drought conditions. In this aspect Kadamalakunta is different from other studies. At the same time, it is to be pointed out here that upper jati people prefer selling some of their assets or mortgage gold ornaments to seasonal migration because it results in loss of prestige and social status. That is, seasonal migration becomes the ultimate resort for the upper caste people for their survival.

Consequently, there is marked change in the patron and client relations in the sense that social bonds are not as strong as they used to be. Also, most of the traditional jatis had to diversify their occupations. For instance, Kuramara, Kuruva, Boya and Vaddera jatis took up occupations which were unknown to then earlier. Indeed drawing upon CPE's in scarcity situations, that has been focused in the recent micro studies (Jodha 1986; Dasgupta 1987), could be observed even in the study village.

Findings from the study village also indicate that drought-relief reassures undertaken by government during droughts served very little purpose and the kind of thrust incorporated in their policies do not reflect in the implementation, which is pointed out in some other studies also (Chen 1987). Long-term policies are neglected, once there is an improvement in the situation. Drought and related factors are influencing social fabric of the village. For instance, dominant position of upper jatis is declining and the reasons for it are several. This is partly due to drought, due to the efforts of governmental and NGO's, due to seasonal migratory process. Drought has to be considered along with other dimensions in analyzing the social change in any region/village, because it is multiple factors that alter the structure of the village and not drought alone.

Governmental and NGO's efforts, which received less attention in other anthropological studies were covered in the personal context along with various other stimuli of change. Though, the adaptation from different drought-prone areas exhibit certain similarities, the local and regional differences, and the nature of external influences will lead to a specific blend, and does not allow broad generalizations. Each drought situation is unique and hence responses and adaptations also vary.

Rural households, both rich and poor, have adapted suitably to cushion themselves against the uncertainties of nature. Given the gradual erosion of traditional hereditary relationships, increasing pressure on CPR's, available resources and the absence of viable need-based government programmes, residents of Kadamalakunta have to rely increasingly on their own initiatives in adjusting to shortages. Though radical transformation is not foreseen in the near future, significant change could take place, because of change in inter-jati relations, gradual economic changes, occupational diversification and break down of traditional socio-political authority. Thus, drought is also an agent of social change. Though people suffer, it makes them to be more active and enterprising, since they are on the lookout for new opportunities to overcome the effects of drought.

The following suggestions based on this study may go a long way in mitigating the sufferings from the drought in the study region as well as in other drought-prone areas. There is a greater need for proper co-ordination between long-term and short-term programmes of the government. For instance, quick relief measures taken for the supply of fodder, fuel wood, drinking water and other scarce resources during severe drought years by government should be properly coordinated in continuation with the long-term measures of social forestry, watershed development programme and so on. More specific programmes like construction of check dams, maintenance of livestock in the drought-prone areas that cater to the local needs which induce people's participation are necessary. Similarly, specific programmes to the landless, women, children and other deprived groups are to be taken up since the existing rural development programmes do not address directly to the particular problems of these groups. The drought-relief measures should be routed through village panchayat/ local self-governments for reaching to the most needy in a better way. For example, supply of fodder directly to the villagers instead of dumping at mandal/regional centres, will serve the actual purpose if the task is entrusted to village panchayats. Wages should be paid more in terms of grain as in JRY. In other words, Programmes evolved for drought-prone areas should take into consideration the socio-cultural, economic and ecological situations/conditions, and strategies for their implementation should be evolved in consultation with the people, taking local needs into consideration.

In fact, it can be argued that the foregoing diagnosis of success in crisis management is contingent upon the existence of acute and lasting famine vulnerability in the first place. The disappearance of large-scale famines in India has indeed coexisted with the resilient persistence of mass poverty and hunger. As researchers has emphasized in the early chapters, moreover, this colossal failure is avoidable and, therefore, inadmissible.

There has been increasing pressure on the farmers in terms of meeting the basic social services such as education and health, which are increasingly privatised and have been emerging as a significant part of domestic expenditure needs. A combination of these stress factors have been at the back of the crisis in the farming sector, which has been manifesting into wide spread suicides, particularly from 1997. The agricultural performance slipped much below the national average, leaving the state in a serious crisis. Even leaving aside comparing with international standards, a comparison with states with in India.

This shows that Andhra Pradesh's agricultural performance is not very impressive. On the other hand, in dry land regions the farmers with relatively low access to irrigation facilities have been getting into deep debt to augment investments in agriculture in the form of tapping ground water resources. This is an unsustainable state for farming to continue as an occupational source. It is no wonder that about 47 percent of farmers in state expressed their willingness to quit farming if given a choice. There was clear warning neo liberal reforms would force adjustment among poor farmers, which without assistance from the state, would intensify their suffering.

Bibliography

Andhra Pradesh District Gazetteers, Anantapur, 1970.

Arnold, David (1984) 'Famines in peasant consciousness and peasant action: Madras 1876-78', In Ranjith Guha (ed.), *Subaltern Studies III: Writings on South Asian History and Society*, Delhi.

Arunachalam, B. (1971) *The Problems of Drought. What crop the Geographer do'r Proceedings*, Symposium on Revenue Development Planning, Madras, 5th-7th January, 1971.

Atkinson (1909) 'Rupees Prices in India, 1870 to 1985', *Journal of Statistical Society*, September, pp. 496-574.

Aykroyd, W.R. (1974) *The Conquest of Famine*, Chatto & Windus, London.

Bhatia, B.M. (1985) *Famines in India: A Study in Some Aspects of the Economic History of India with Special Reference to Food Problem*, Delhi: Konark Publishers.

Bhatia, B.M. (1991) *Famines in India.*

Bhatt, V.V. (1963) *Aspects of Economic Change and Policy in India 1800-1960*, Bombay.

Bhattacharya, Sabyasachi, Sumit Guha, Raman Mahadevan, Sakti Padhi, D. Rajasekhar and G.N. Rao (1919) *The South Indian Economy: Agrarian Change, Industrial Structure and State Policy 1914-1947*, OUP, Delhi.

Blix, G., Y. Hofvander and B. Vahlquist (eds.) (1971) *Famines A*

Symposium Dealing with Nutrition and Relief Operations in Times of Disasters, Almquist and Wikselle, Uppsalel, Sweden.

Bose, Sugatha (ed.) (1994) *Credit Markets and the Agrarian Economy of Colonial India*, OUP, Delhi.

Bowen, H.V. (2002) *Revenue and Reform: The Indian Problem in British Politics 1757-1773*, Cambridge: Cambridge University Press.

Brook, D.A. Wash (1978) 'Economic Development and Social Stratification in Rural Madras: The Dry Region, 1878-1929', In G. Dewey and A.G. Hopkins (eds.), *The Imperial Impact: Studies in the Economic History of Africa and India*, The Athlene Press.

Chandra, Bipan (1969) *The Rise and Growth of Economic Nationalism in India*, New Delhi.

Charlesworth, Neil (1982) *British Rule and the Indian Economy 1800-1984*, Macmillan, London.

Collier, Michael and Webb, Robert H. (2002) *Floods, Droughts, and Climate Change*, Tucson: University of Arizona Press.

Cornish, W.R. (1871) *Report of the Census of the Madras Presidency 1871*, Vol. 1.

Currey, B. and G. Hugo (1984) 'Famine as a Geographical Phenomenon', *Geo Journal Library*, Boston: D. Reidel.

Dalyell, R.A. (1867) *Memorandum on the Madras Famine of 1866*, Madras.

Das, Durga (1969) *India from Curzon to Nehru and After*, London.

Davis, Mike (2001) *Late Victorian Holocausts: El Nino Famine and the Making of the Third World*, New York: Verso Books.

Desai, Meghnad, Tapan Raychaudhuri and Dharma Kumar (1983) *The Cambridge Economic History of India*, Volume 2, Cambridge: Cambridge University Press.

Devereux, S. (2000) *Famine in the twentieth century*, IDS Working Paper 105, Brighton: Institute of Development Studies.

deWaal, A. (1989) *Famine that Kills, Darfur, Sudan, 1984 - 85*, Oxford: Clarendon Press.

Digby, William (1878) *The Famine Campaign in Southern India:*

Madras and Bombay Presidencies and Province of Mysore, 1876-1878, vol. 1.

Digby, William (1901) *Prosperous British India*, T. Fisher Unwin.

Dreze, J. and A.K. Sen (1989) *Hunger and Public Action*, Oxford: Clarendon Press.

Dreze, Jean (1991), 'Famine Prevention in India', In Drèze, Jean and Sen, Amartya (eds.), *The Political Economy of Hunger: Famine Prevention*, Oxford: Oxford University Press.

Dutt, R.C. () *Famines in India*, Appendix O, p. 197.

Dutt, R.C. (1902) *Economic History of India*, Publication Division, Government of India.

Edwardes, Michael (1963) *The Last Years of British Rule in India*, London.

Encyclopedia of Britannica (A Survey of Universal Knowledge), vol. 7-9.

Epstein, T. Scarlet (1962) *Economic Development and Social Change in South India*, Oxford University Press, Bombay.

Famine Commission (1898) Report, Chapter II, para 20.

Famine Commission Report (1880), Vol. III, Appendix 1, "Conditions of the Country and the People".

Francis, W. (1905) *Anantapur District Gazetteer*.

Ganguli, B.N. (1938) *Trends of Agriculture in the Ganges Valley: A Study in Agricultural Economics*, pp. 288-289.

Ghose, Ajit Kumar (1982) 'Food Supply and Starvation: A Study of Famines with Reference to the Indian Sub-Continent', *Oxford Economic Papers*, Oxford: Oxford University Press, 34(2): 380.

Gilmour, David (2007) *The Ruling Caste: Imperial Lives in the Victorian Raj*, New York: Macmillan.

Goodall, Frederick () 'Famine in Mysore', *Calcutta Review*, LXXXVI: 261-76.

Gourlay, J. (2003) *Florence Nightingale and the Health of the Raj, Aldershot, Hants*, England; Burlington, VT: Ashgate.

Habib, Irfan (1984) 'Studying a Colonial Economy without Perceiving Colonialism', *Social Scientist*, vol. 12, no. 12, December.

Henvey (1871) *Narrative of Famine in NWP.*

Howe, P. (2002) 'Reconsidering Famine', *IDS Bulletin*, 33(4).

Hunter, W.W. (1885) *The Imperial Gazetteer of India*, Vol. 1, London.

Indian Famine Commission 1898, Appendices.

James, Lawrence (2000), *Raj: The Making and Unmaking of British India*, Macmillan.

Jha, Manoranjan (1972) *Role of the Central Legislature in the Freedom Struggle*, New Delhi.

Knowles, L.C.A. (1950) *The Industrial and Commercial Revolutions in Great Britain during the Nineteenth Century*, London, Reprint.

Kumar, Dharma (1978) *Land and Caste in South India*, Manohar Publishers, New Delhi.

Lardinois, Roland (1985) *Famine Epidemics and Mortality in South India – A Reappraisal of the Demographic Crisis of 1876-78*, vol. 11.

Loveday (1914) *History and Economics of Indian Famines*, London: G. Bell and Sons Ltd.

Macleane, C.B. (1979) *Standing Information Regarding the Administration of the Madras Presidency*, Madras.

Madras Famine Review, Appendix 'P', quoted in Famine Histories.

Malceane, C.D. (1879) *Standing Information Regarding the Administration of the Madras Presidency*, Madras.

Mander, Harsh (2009) 'Barefoot: Colonial Legacy of Famine Codes', Chennai: *The Hindu*, retrieved 3rd October 2010.

Mann, H. (1955) *Rainfall and Famine (A Study of Rainfall in the Bombay Deccan, 1865-1938)*, Bombay.

Mathews, David Hall (1999) 'Colonial Ideologies of the Market and Famine Policy in Ahmednagar district, Bombay Presidency 1870-84', *Indian Economic and Social History Review*, XXXVI(3).

McIver, Lewis (1883) *The Imperial Census of 1861 — Madras Presidency*, Madras.

Meteorological Glossary, Meteorological Off., London, 1957.

Minchin, J.I. (1864) *Collector of Bellary to the Board of Revenue*, Madras, 12th May.

Mukherji, Nilamani (1964) 'The Impact of the Ryotwari System on the Economy of the Madras Presidency.' In B.N. Ganguli (ed.), *Readings in Indian Economic History,* Delhi.

Munro Sir Thomas, *Minutes, Letters and Dispatches on Food scarcity in Arbuthnot,* Vol. 11, p. 215.

Nadikarni, M.V. (1985) *Socio-economic Conditions in Drought-Prone Areas,* Concept Publishing Co., New Delhi.

Nash, J. Madeleine (2003) *El Niño: Unlocking the Secrets of the Master Weather-Maker,* London: Warner Books.

Nash, Vaughan (1901) *Great Famine.*

Olsen, K.W. (1987) 'Man-made Drought in Rayalaseema', *Economic and Political Weekly,* XXII(11): 441-43.

Pandian, M.S.S. (1990) *The Political Economy of Agrarian Change: Nanchilnadu 1880-1939,* Sage Publications, New Delhi.

Patel, S. (1952) *Agricultural Labor in India and Pakistan,* Bombay: Current Book House.

Pratap Kanakalapati (2015) 'Famines and Agrarian Conditions in Anantapur District : A Historical Review', *Andhra Pradesh History Congress Proceedings,* p. 207.

Proceedings of the Board of Revenue.

Raghavaiyangar, S. Srinivasa, *Memorandum on the Progress of the Madras Residency,* Appendix, Section V.

Raju, A. Sarada (1941) *Economic conditions in the Madras Presidency 1800.*

Raju, G. Suseela (1990) 'Critical Analysis on the Impact of Moneylenders on Peasants in Andhra Pradesh (1900-1947)', *ITIHAS,* XVI(2).

Ravallion, Martin (1987) 'Trade and Stabilisation: Another Look at British India's Controversial Foodgrain Exports', *Explorations in Economic History,* 24(4): 354-370.

Reddy, V. Ratna (1900) *Irrigation in Colonial India – A Study of Madras Presidency during 1860-1900.*

Report of the Commission of Famine in Bengal, Bihar and Orissa, 1867, part III, para 63.

Report of the Indian Famine Commission, 1898, Vol. 1.

Report on Effect of Famine on Manchester Trade, dated 8th May 1861, para 31.

Revenue Consultations, Madras, 4th August 1861, (A.P.S.A.).

Revenue dispatches to England, 1821-1826, Vol. 60088 (T.N.A.).

Ricardo, David (1815) *An Essay on the Influence of a Law Price of Corn on profits of Stock*, Vol. IV.

Rothermund, Dietmar (1988) *Economic History of India from pre-colonial times to 1986*, Manohar Publications, New Delhi.

Roy, Tirthankar (1999) *Traditional Industry in the Economy of Colonial India*, Cambridge University Press.

Roy, Tirthankar (2006),] *The Economic History of India, 1857-1947*.

Sahu, A.C. (1978) *Export of Food Grains: A Potent cause of famine in India between 1860-1900*, IHC proceedings, Modern India section, 39th session Hyderabad, vol. II.

Sahu, A.C. (1980) *Diversion of arable lands from food crops to exportable cash crops: One of the causes of frequent famines in India between 1860-1900*, IHC proceedings, modern India section, 41st session Bombay.

Satyanarayana, A. (1990) *Andhra Peasants under British Rule: Agrarian Relations and Rural Economy 1900-1940*, Manohar Publishers, New Delhi.

Sen, Amartya (1983) *Poverty and Famines : An Essay on Entitlement and Deprivation*, Oxford: Clarendon Press.

Singh, N. (2002) *Population and Poverty*.

Singh, Tapeshwar (1978) *Drought-Prone Areas in India*, People's Publishing House, Delhi.

Sir James, A. Murray (ed.), *Palgrave's Dictionary of Political Economy*.

Smith, Baird (1861) *Effects of Famine on Manchester Import Trade in North Western Provinces*.

Southward, 'Famine', *Encyclopedia of Social Sciences*, Vol. VI.

Srivastava, H.C. (1968) *The History of Indian Famines from 1858-1918*, Agra: Sri Ram Mehra and Co.

Statistical Atlas of India (1895).

Stokes, Eric (1959) *The English Utilitarian and India,* Oxford.

Sudhakar, Pasala (2006) 'Drought Distress Suicide of Farmers in Anantapur district', In P. Prasangi (ed.), *Osmania Journal of Historical Research Bi-Annual Journal*, VII(6): 20-37.

The Irrigation Commission Report, 1972.

The Land System of India, Vol. LXXV (1863).

The Wealth of Nations, Book I, Chapter VIII (Modern Library Editing).

Thomas, P.J. (1926) *Mercantilism and East India Trade,* London.

Thomas, P.J. and B. Natarajan (1933) *Economic Depression in the Madras Presidency, 1820-1854,* Madras.

Thorburn, S.S. (1808) *Musalmans and Money-lenders in the Punjab Provinces*, para 67.

Visaria, Leela and Pravin Visaria (1983) 'Chapter V: Population (1757-1947)', In Dharma Kumar and Meghnad Desai, *The Cambridge Economic History of India*, Cambridge: Cambridge University Press.

Walford (1878) 'Famines of the World: Past and Present', *Journal of Statistical Society*, XLI: 433-526.

Walsh, Judith E. (2006) *A Brief History of India*.

Washbrook, David (1994) 'The Commercialization of Agriculture in Colonial India: Production, Subsistence and Reproduction in the Dry South, c. 1870-1930', *Modern Asian Studies*, 28(1).

Wittfogel, A. Karl (1957) *Oriental Despotism: A Comparative Study of Total Power*.

Wong, J.W. (1998) *Deadly Dreams: Opium, imperialism, and the Arrow War (1856–1860) in China*, Cambridge: Cambridge University Press.

Index

The fate of paddy land at Kanaganipalli village

A depleted tank at Gooty taluk

A farmer in utter distress in his ground-nut field

An orange garden at Thadimarri mandal

Dried borewell in Kanaganipalli village

Loss of groundnut crop at Gangampalli, Ramagiri mandal

Sheep grazing after failure of monsoon in horse gram field